ENCYCLOPAEDIA OF ENGINEERING PHYSICS - III

ELECTROMAGNETICS

By

Dr. Shalender Singh

DISCOVERY PUBLISHING HOUSE PVT. LTD.

NEW DELHI-110 002

Published by:
Namit Wasan

DISCOVERY PUBLISHING HOUSE PVT. LTD.
4383/4B, Ansari Road, Darya Ganj
New Delhi-110 002 (India)
Phone : +91-11-23279245; 23253475; 43596065
E-mail : discoverybooksindia@gmail.com
discoverypublishinghouse@gmail.com
namitwasan9@gmail.com
web : www.discoverypublishinggroup.com

Edition: **2020**

ISBN: 978-81-8356-349-9 (Set)

ISBN: 978-81-8356-416-8

Electromagnetics

Printed at:
Infinity Imaging Systems
Delhi

Preface

The present book "Electromagnetics" meant for engineering students. The present book is an attempt to fulfil the need of all engineering students of U.P.T.U. as well as for engineering students of other states. It covers the complete syllabus of physics prescribed by Technical Universities. The treatment given in simple, lucid and comprehensive. The language used is simple.

Though nothing can be claimed as original but the subject matter has been arranged in my own style.

Suggestion for improvement of the book will be greatly acknowledged.

Author

CONTENTS

1

Electromagnetic Induction

INTRODUCTION

In 1864 *James Clerk Maxwell* brought together and extended four basic laws in electromagnetism such as. *Gauss's law in electrostatics, Gauss's law in magnetism. Ampere's law and Faraday's law*. In fact entire theory of the electromagnetic field is condensed into these four laws. These laws govern the interaction of bodies which are magnetic or electrically charged or both. A complete set of relations giving the connection between the charges at rest (electrostatics), charges in motion (current electricity), electric fields and magnetic fields (electromagnetism) were derived theoretically and summarized in four equations by Maxwell, called *Maxwell's equations*. Maxwell's equations are based on the assumption that matter has atomicity structure and is continuous. These equations are satisfied by the electric and magnetic components of the electromagnetic field in all cases.

Like Newton's law in Mechanics, Maxwell's equations are the back bone of electrodynamics. The success of these equations is remarkable, it includes the fundamental operating principles of all large-scale electromagnetic devices such as motors, electronic computers, microprocessors, cyclotron, television, microwave radar, transmitting system, receiving system etc. Classical electromagnetic theory constituted by Maxwell's equations predicts that accelerated charges produce electromagnetic waves which propagates in space with a velocity equal to the velocity of light. It also showed that light waves are electromagnetic in character. In 1888 *Heinrich Hertz* produced and detected these waves. This firmly established wave theory of light.

Maxwell's equations are used in integral as well as in differential form depending upon the problems to be solved. Like others. Maxwell's equations have certain limitations. They most successfully explain electromagnetic interaction between large aggregates of charges such as radiating antennas, electric circuits etc. but the electromagnetic interactions between fundamental particles could not be correctly explained on the basis of results derived from the Maxwell's equations. They are treated according to the laws of quantum mechanics by a technique called quantum electrodynamics.

AMPERE'S CIRCUITAL LAW

Ampere's circuital law in magnetostatics is analogous to Gauss's law in electrostatics. It states that *"the line integral of magnetic field intensity* $\vec{H}$ *about any closed path is exactly equal to the net current enclosed by that path,"* that is,

$$\oint \vec{H}.d\vec{l} = I \qquad ...(1)$$

The positive current is taken in the direction of advancement of right handed screw turned in the direction in which the closed path is traversed (Fig. 1.1a).

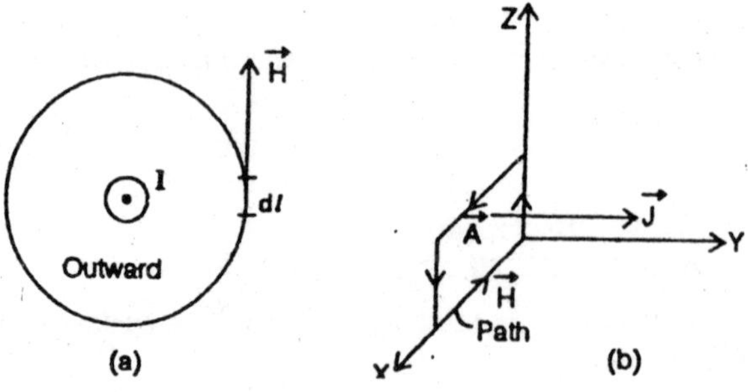

Fig. 1.1

For the more general case, where the current density is non-uniform we have

$$\oint \vec{H}.d\vec{l} = I = \int_s \vec{J}.d\vec{S} \qquad ...(2)$$

where $\int \vec{J}.d\vec{S}$ is the total current passing through tee closed curve.

Eqn. (1) is also an integral form of Ampere's law. The law can be applied inside or outside a conductor, but total current enclosed by the path should be known.

Ampere's circuital law is frequently used to calculate $\vec{H}$ in highly symmetric situation. The standard current configurations which can be easily handled by Ampere's law are, infinite straight lines, infinite planes, infinite solenoids, toroids etc.

Usually, following two conditions must be fulfilled for the accurate determination of $\vec{H}$ by using Ampere's circuital law.

1. At each point on the closed path, magnetic field intensity $\vec{H}$ must be either tangential or normal to the path.
2. In the case where $\vec{H}$ is tangential to the path, it must be equal at all points of the path.

In term of magnetic induction vector $\vec{B}$, Ampere's circuital law states that *the line integral of* magnetic induction vector $\vec{B}$ around any closed path is equal to μ_0 (permeability of free space) times the total current crossing the surface bounded by the closed path. That is,

$$\oint \vec{B}.d\vec{l} = \mu_0 I$$

In differential form, Ampere's law may be expressed as

$$\text{Curl } \vec{B} = \mu_0 \vec{J} \quad \text{or} \quad \text{curl } \vec{H} = \vec{J}$$

To prove the law, let us consider the field intensity at a point P distant r from a long straight filamentry wire carrying a current r as shown in Fig. 1.2. Now consider the circular path of radius r centred on this current carrying wire. The field intensity $\vec{H}$ any point P on the circular path is

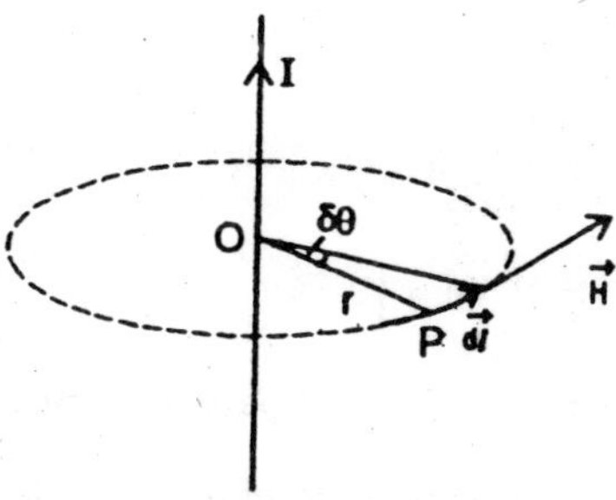

Fig. 1.2

$$H = \frac{I}{2\pi r} \qquad ...(1)$$

For every point on the circular path of radius r the field intensity has the same magnitude as given by eqn. (1) and is parallel to the tangent to the path. Therefore, the line integral of field intensity $\vec{H}$ around the circular path centred on the-current carrying wire is given by,

$$\oint \vec{H}.d\vec{l} = \oint H\,dl = \frac{I}{2\pi r}\oint dl$$

$$= \frac{I}{2\pi r} \times 2\pi r = I$$

Thus, we have $\oint \vec{H}.d\vec{l} = I$

APPLICATIONS OF AMPERE'S LAW

Magnetic Field Due to a Long Straight Current Carrying Filamentry Wire

Let us consider a long straight filamentry wire carrying a steady current I (Fig. 1.3). From the symmetry of the wire it is obvious that the magnetic lines of force are concentric circles centred on the wire. In order to find the magnetic field $\vec{B}$ at a point P, we draw a circle passing through the point P with centre at O on the wire. The radius of the circle is OP = r. By symmetry, all points on the circle are equivalent and hence, the magnitude of field $\vec{B}$ should be same at all these points and is directed tangentially to the circle everywhere. Moreover, $\vec{B}$ and $d\vec{l}$ arc always directed along the same direction, therefore line integral of $\vec{B}$ along the boundary of the circular path is

$$\oint \vec{B}.d\vec{l} = \oint B\,dl = B\oint dl$$

or $$\oint \vec{B}.d\vec{l} = B.2\pi r$$

From Ampere's law, we have

$$\oint \vec{B}.d\vec{l} = \mu_0 I$$

$\therefore$ $$B.2\pi r = \mu_0 I$$

or $$B = \frac{\mu_0 I}{2\pi r}$$

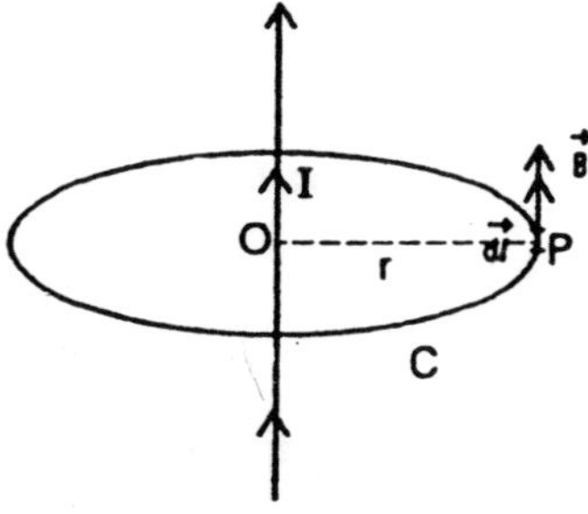

Fig. 1.3

Magnetic Field Induction of a Solenoid

A solenoid is a coil of wire wound on the surface of a hollow cylinder of a card-board or china-clay and has a length very large compared to its diameter (Fig. 1.4). It is experimentally observed that when a current passes through such a solenoid the magnetic field outside is very small compared with that inside. Further the line of induction inside the solenoid are straight and parallel showing a uniform magnetic field except near the edges (Fig. 1.5).

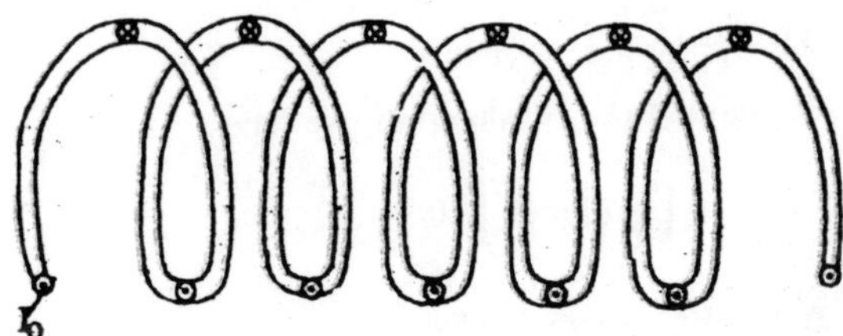

Fig. 1.4

To determine the magnetic field at a point inside a current carrying, solenoid consider a closed rectangular path abed in which the side ab is parallel to the axis and sides be and da are very long so that the side cd is far from the solenoid (Fig. 1.5) and thus the field at this side is negligible. As the solenoid is long and the rectangle is not too near the either ends, the field is at right angles to the sides be and da.

Thus, the line integral of $\vec{B}$ along the closed rectangle abed is

$$\oint \vec{B}.d\vec{l} = \int_a^b \vec{B}.d\vec{l} + \int_b^c \vec{B}.d\vec{l} + \int_c^d \vec{B}.d\vec{l} + \int_d^a \vec{B}.d\vec{l} \quad ...(1)$$

Now, $\int_b^c \vec{B}.d\vec{l} = \int_d^a \vec{B}.d\vec{l} = 0$

($\because$ along bc and da, $\vec{B}.d\vec{l}$ = Bdl cos 90° = 0

and $\int_c^d \vec{B}.d\vec{l} = 0$

($\because$ $\vec{B}$ is zero far outside the solenoid)

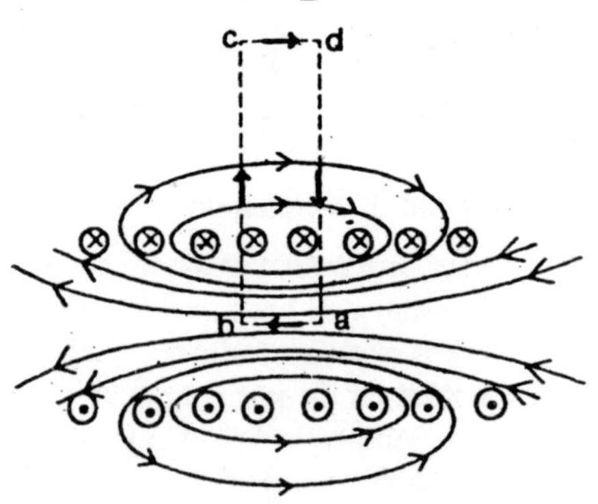

Fig. 1.5

$$\therefore \quad \oint \vec{B}.d\vec{l} = \int_a^b \vec{B}.d\vec{l} \qquad ...(2)$$

As $\vec{B}$ is parallel to $d\vec{l}$ along ab, we have

$$\int_a^b \vec{B}.d\vec{l} = \int_a^b B\,dl = B\int_a^b dl$$

or $\int_a^b \vec{B}.d\vec{l} = Bl$, where l = ba

Therefore, eqn. (2) becomes

$$\oint_{abcd} \vec{B}.d\vec{l} = Bl \qquad ...(3)$$

Applying Ampere's circuital law to the closed rectangular path abed, we have

$$\oint \vec{B}.d\vec{l} = \mu_0 I \qquad ...(4)$$

If n be the number of turns per unit length along the length of the solenoid, then nl turns cross the rectangle abed. Each turn carries a current I_0. Therefore, the net current, crossing the rectangle abed equals to nlI_0

$$\therefore \qquad \oint \vec{B}.d\vec{l} = \mu_0 n l I_0 \qquad ...(5)$$

Comparison of eqns. (3) and (5) yields,

$$Bl = \mu_0 \, nl \, I_0$$

or

$$B = \mu_0 \, n \, I_0 \qquad ...(6)$$

It is clear from eqn. (6) that the field $\vec{B}$ is independent of the length and diameter of the solenoid and is uniform over the cross-section of the solenoid.

If the solenoid is wrapped on a core of material of permeability μ or relative permeability μ_r. Then

$$B = \mu n \, I_0 = \mu_0 \mu_r n I_0$$

Magnetic Field of a Toroid (or Endless solenoid)

If the a solenoid is bent round in the form of a closed ring and their ends are joined, we get a toroid. Toroid may also be constructed by closely winding uniform turns of wire over a non-conducting ring. The magnetic field in such a toroid may be obtained by using Ampere's circuital law.

In order to find the magnetic field at a point P inside the toroid carrying a current I_0 at a distance r from the centre O, we draw a circle through the point P concentric with the toroid as shown in Fig. 1.6. By symmetry the direction of the field $\vec{B}$ is everywhere tangential to the circle and the magnitude of $\vec{B}$ is same at all points on the circle.

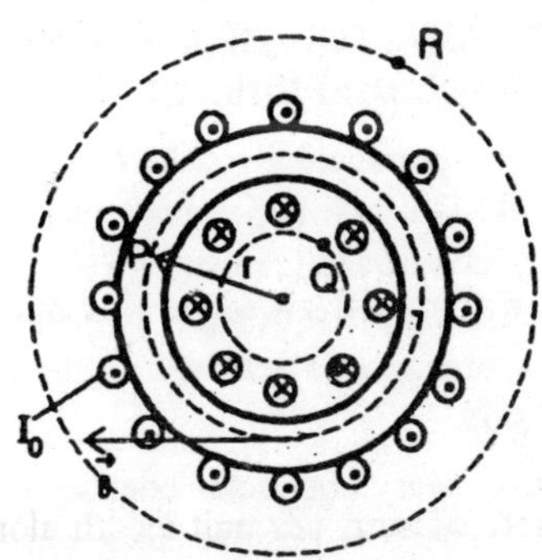

Fig. 1.6

Therefore, the line integral of magnetic field $\vec{B}$ along the circular path of radius r is

$$\oint \vec{B}.d\vec{l} = \int B.dl = B\int dl = B(2\pi r)$$

If the total number of turns in the toroid is N, then the current crossing the area bounded by the circle is Wo.

Applying Ampere's circuital law on the circle of radius r, we have

$$\oint \vec{B}.d\vec{l} = \mu_0 NI_0$$

$$2\pi r\ B = \mu_0 NI_0$$

or $$B = \frac{\mu_0 NI_0}{2\pi r}$$

Thus, the magnitude of magnetic induction $\vec{B}$ varies inversely with r. If *l* be the mean circumference of the toroid, then $l = 2\pi r$, therefore,

$$B = \frac{\mu_0 NI_0}{l}$$

The magnetic field at a point outside the toroid such as R is zero, because each turn of winding passes twice the area enclosed by the circle through R, carrying equal currents in opposite directions. So the total current within the circle is zero.

Similarly, at an internal point, like Q the field is also zero because the circle passing through Q encloses no current. Thus, the field at* toroid is confined only within the core.

MOTION OF A CONDUCTOR IN A MAGNETIC FIELD—INDUCED POTENTIAL DIFFERENCE

(i) Suppose a thin conducting rod MN of length *l* is placed in a magnetic field B which is perpendicular to the plane of paper and directed downward [shown by (×) marks]. Let the rod is being moved in the plane of paper perpendicular to its own axis and direction of magnetic field with a constant velocity v towards right hand side.

(ii) We know that every conductor consists of some positive and some negative charges – positive charges in the form of positively charged ions while negative charges in the form of electrons, of these only the electrons are capable of moving freely in the conductor. When rod moves with velocity v, then charges present in it a also move with the same velocity v, in the magnetic field B towards right hand side. We have read that a charge q moving in a magnetic field B with some velocity v perpendicular to the field, experience a magnetic force F_m given by

$$F_m = qvB$$

(iii) According to Right hand palm rule no. 2, Lorentz force F_m on the positive charge will be directed towards end M while that on negative charge towards end N. But because only the electrons are free to move, they start moving towards the end N. As a result of this, end N becomes negatively charged (due to excess of electrons) while end M gets equally positively charged (due to deficiency of electrons). Thus an electric potential difference is produced between two ends of rod due to which an electric field is produced within the rod along its length in downward direction. If V is the potential difference produced across a conductor of length l, then

$$E = V/l \qquad ...(1)$$

(iv) Due to this electric field an electric force also acts on each charge q in the conductor whose magnitude is

$$F_e = qE.$$

Direction of force F_e on positive charge will be towards end N while on negative charge towards end M. Thus we find that direction of electric force on each charge is opposite to that of Lorentz force F_m.

(v) Due to continuous motion of the conductor, as more and more electrons reach the end N electric force F_e goes on increasing till it becomes equal to Lorentz force F_m (Lorentz Force is fixed from the very beginning). In this situation, resultant force on each charge in the rod becomes zero and movement of electrons towards end N gets stopped *i.e.,* potential difference produced between two ends of rod, which was increasing till now, also becomes constant.

(vi) Thus, when potential difference across the rod becomes constant

$$F_e = F_m \text{ or } qE = qvB \text{ or } E = vB \qquad ...(2)$$

From eq. (1), we find, $V = El = vBl$ ..(3)

LORENTZ FORCE AND FARADAY'S LAW OF ELECTROMAGNETIC INDUCTION

(i) According to Faraday's experiments we have seen that whenever magnetic flux linked with a closed circuit changes, an electric current is induced in the circuit. This can be understood in the basis of Lorentz Force.

(ii) Suppose a conducting rod MN of length l is allowed to move with a velocity v towards right hand side, on the two rails of a U-shaped conductor PQRS. Let the U-shaped conductor is placed in a magnetic field B which is perpendicular to the plane of paper and directed downward. As the conducting rod moves, free electrons present in it experience a Lorentz force which is directed along the length of rod towards its end N. Due to this force, electrons are shifted from upper end M to lower end N. When they reach the end N, they find a closed path NRQM and thus move along the path N $\rightarrow$ R $\rightarrow$ Q $\rightarrow$ M. Because the direction of current in a circuit is taken opposite to direction of motion of electrons, hence a current flows in the circuit in anticlockwise direction. This current continues to flow in the circuit till the rod MN remains in motion. This shows that *an emf is induced across the moving rod which is responsible for maintaining current in the circuit.* As this emf is produced due to the motion of rod, hence it is also called as motional emf. It is now clear from above discussion that *basic cause for generation of this induced emf or current is the Lorentz Forces acting on free electrons in the moving rod.*

(iii) As the conducting rod moves towards right hand side, magnetic flux linked with the closed circuit MQRN also increases. This just verifies Faraday's concept that induced current continues to flow in the circuit so long as magnetic flux linked with it changes.

(iv) Let e is the induced emf produced in the moving rod MN and i the induced current flowing in the circuit due to this induced emf. We have read that a current carrying conductor placed in a magnetic field always experience a force whose direction is given by Right hand palm rule no. 2. Hence force exerted on the current–carrying rod MN of length l due to magnetic field B, is given by

$$F' = ilB$$

According to right hand palm rule no. 2, this force will be in the plane of paper, perpendicuiar to conducting rod and directed towards left hand side *i.e.,* this force opposes the motion of conducting rod towards right hand side (This happens according to Lenz's rule). Hence to keep the rod moving with constant velocity v towards right hand side (to maintain the same induced current in same direction), a force F equal and opposite to F' is to be applied on it *i.e.,*

$$F = -F' = -ilB$$

(v) Let the rod MN moves a small distances Δx in time interval Δt under the force F and acquires new position M'N'. Then the work done on the moving rod is given by

$$W = F\Delta x = -ilB.\Delta x = -ilB.(v\Delta t) = -vlB.q$$

[as $(i \times \Delta t) = q$ = charge flown in the circuit in time Δt].

This work provides the necessary energy for flow of charge q in he circuit.

(vi) We know that energy supplied by a cell for flow of unit charge throughout the whole circuit is known as emf of the cell. Therefore induced emf produced in the moving rod MN, which is acting just like a cell in the close circuit MQRN, is given by

$$e = W/q = -Bvl \qquad ...(1)$$

(vii) Now in the interval Δt, area of the closed circuit increases from MNRQ to M'N'RQ. Hence change in magnetic flux linked with the circuit in time interval Δt is given by

$\Delta\phi$ = magnetic field × change in area = B $(l\Delta x)$

Hence the rate of change of magnetic flux is

$$\frac{\Delta\phi}{\Delta t} = Bl\,\frac{\Delta x}{\Delta t} = Blv \quad ...(2) \qquad \left[v = \frac{\Delta t}{\Delta t} = \text{velocity of rod}\right]$$

Comparison of eq. (1) and (2) yields $e = -\Delta\phi/\Delta t$

when $\Delta t \to 0$, $e = -\left(\frac{d\phi}{dt}\right)$

This is *Faraday's Law of electromagnetic induction.*

(viii) The negative sign shows that induced current always flows in such a direction so as to oppose the cause of its production. This fact is always verified, whatsoever be the cause of generation of induced current we may understand. If induced current is supposed to be generated due to the motion of conducting rod, then direction of induced current in the circuit is such that (anticlockwise) the *force exerted on the rod opposes the motion of the rod.* On the other hand, if the cause of generation of induced current is taken as increase in magnetic flux, then direction of current in the circuit is such that the magnetic field produced to this current is just opposite to the original field B *i.e., it opposes the increase in magnetic flux.*

(ix) **Direction of Induced Current : Flemming's Right Hand Rule**—If we stretch the thumb, forefinger and the central finger of our right hand mutually perpendicular, and the fore finger points in the direction of magnetic field and them in the direction of motion of the conductor, then the central finger will give the direction of induced current produced in the conductor.

MUTUAL INDUCTION

(i) Suppose two coils P and S are placed close to each other. A battery and a tapping key K are connected in the P–coil while a galvanometer is connected across the S–coil. The coils P and S are known as PRIMARY and *Secondary* coils respectively.

(ii) As soon as key K is pressed, current in the primary coil starts growing up and a magnetic flux starts building up with the coil. As a result of this an induced emf is produced in the coil itself due to self induction which opposes the growth of current in the primary coil. At the same time, because two coils are placed near to each other, hence a magnetic flux also starts building up in the secondary–coil *i.e.*, an induced emf and therefore an induced current is also produced in the secondary coil. This emf also opposes the growth of current in primary coil.

(iii) Similarly, when key K is opened, again induced emf is developed both in primary and secondary coils. But now the sign of induced emf will be opposite to the one, which is produced when key was pressed. Thus, during both the growth and decay of current in the primary coil, an opposing induced emf is produced in the nearby secondary coil.

(iv) Thus, *the phenomenon of electromagnetic induction in which, on changing the current in one coil an opposing induced emf or induced current is produced in a neighbouring coil, is known as Mutual Induction.*

(vi) **Coefficient of Mutual Induction :** Suppose I is the current flowing through the primary coil at any instant. It is found that the magnetic flux ϕ linked with the S-coil at that instant will be directly proportional to the current passing through P-coil at that instant *i.e.,*

$$\phi \propto I \text{ or } \phi = MI \qquad ...(1)$$

where M is called as *Coefficient of Mutual Induction* or simply *Mutual Inductance* of the two coils.

If $I = 1$, then $f = M$ or $M = \phi$.

i.e., Mutual inductance of the two coils is numerically equal to the magnetic flux linked with one coil when a unit current flows through the neighbouring coil.

(vii) The induced emf produced in the secondary coil due to changing magnetic flux is given by

$$e = -\frac{d\phi}{dt} = -\frac{d}{dt}(MI) = -M\left(\frac{dI}{dt}\right) \qquad ...(2)$$

Hence, (dI/dt) represents the rate of change of current in the primary coil and e is the induced emf produced in S–coil due to mutual induction. The negative sign shows the opposing nature of the induced emf. If we consider the magnitude only then from eq. (2), we get

$$|e| = M\frac{dI}{dt} \quad \textit{i.e.}, \text{ If } \frac{dI}{dt} = 1, \text{ then } M = |e|$$

i.e., Coefficient of mutual inductance of two coils is numerically equal to the induced emf produced in one coil, when the rate of change of current is unity in the neighbouring coil.

(viii) In SI system of units, the unit of mutual inductance is also HENRY (H)

Again $M = \dfrac{|e|}{(dI/dt)}$ *i.e.,* If $|e| = 1$ volt

and $(dI/dt) = 1$ ampere/second, then

$$M = \frac{1 \text{ volt}}{1 \text{ amp/sec}} = 1 \text{ henry}$$

i.e., Mutual inductance of two coils is equal to one henry, if a rate of change of current of 1 ampere per second in one coil induces an emf of one volt in the neighbouring coil.

EDDY CURRENTS

(i) According to Faraday's Experiments we know that when magnetic flux linked with a closed circuit changes, an induced current is produced in the circuit. Foucault observed that if a metallic plate is placed magnetic flux linked with it changes continuously with respect to time, then also induced currents are produced in the whole volume of the metal. These currents oppose the motion of metallic plate or oppose the change in magnetic flux linked with the plates. These currents are called as Eddy currents.

(ii) Suppose a copper plate is placed in a magnetic field which is perpendicular to plate and directed downward. When we draw the plate out of magnetic field, we feel some opposing force. It is due to the reason that on withdrawing the plate, magnetic flux linked with the plate decrease. As a result of this decrease in flux, induced current loops are produced in the plate coming out of the magnetic field. According to Lenz's Rule, the direction of induced current is such that it opposes the decrease in magnetic flux (Induced current produces a magnetic field in the same direction as that of original magnetic field) *i.e.*, it opposes the motion of plate out of magnetic field.

(iii) Similarly, when we insert the plate into magnetic field, then eddy currents are produced just in opposite direction so that they may now oppose the increase in magnetic flux or motion of plate into the magnetic field.

(iv) **Undesirable effect of eddy current :** The production of eddy currents results in the loss of energy in the form of heat. Hence there occurs a huge loss of energy in the cores of armature–coils of dynamo and motors, and in the frame of transformers due to production of eddy currents. To minimise this loss of energy, the core or the frame is not taken in the form of a single piece of soft iron, but it is prepared after joining a many thin laminas of soft iron by varnish. The core so obtained is called as laminated soft iron core. A laminated core offers a high resistance to eddy currents which therefore become quite weak. Thus by using a laminated soft iron core, losses due to production of eddy currents are minimized.

GROWTH AND DECAY OF CURRENT IN L–R CIRCUIT

(a) Growth of current in L–R circuit

The circuit shown in figure consists of an inductance L and a resistance R connected to a battery of emf E and a key K. When the key is closed, the current does not attain the value (E/R) immediately because of self inductance of the coil [self inductance of the coil induces a current in the circuit in a direction opposite to that of main current]. If (dI/dt) is the rate of change of current in the coil at any instant, then self induced emf V_L produced in the coil is given by

$$V_L = -L(dI/dt)$$

According to Ohm's Law, we can now write

$$E - L\,(dI/dt) = RI \text{ or } RI - E = -L\,(dI/dt)$$

$$\text{or } \frac{dI}{RI - E} = -\frac{I}{L}dt$$

$$\text{or } \frac{dI}{I - (E/R)} = -\left(\frac{R}{L}\right)dt \qquad ...(1)$$

Integrating and considering that at t = 0, I = 0

$$\int_0^I \frac{dI}{I - (E/R)} = -\frac{R}{L}\int_0^t dt$$

$$\text{or } \text{Log}\left(I - \frac{E}{R}\right) - \text{Log}\left(-\frac{E}{R}\right) = -\left(\frac{R}{L}\right)t$$

$$\text{or } \text{Log}\left(I - \frac{E}{R}\right) = \text{Log}\left(-\frac{E}{R}\right) + \text{Log}_e^{-(R/L)t}$$

$$\text{or } I - \frac{E}{R} = -\frac{E}{R}\,e^{-(R/L)t}$$

$$\text{or } I = \frac{E}{R}\,(1 - e^{-(R/L)t}) = I_0\,[1 - e^{-(R/L)t}] \qquad ...(2)$$

where I_0 = (E/R) = steady or maximum value of current.

Production of Alternating Current (A.C. Generator)

(i) A device which is used to convert mechanical energy into electrical energy is called an electric generator.

(ii) **Principle :** It is based on the principle of electromagnetic induction. When a coil is rotated rapidly in a strong magnetic field about an axis perpendicular to the direction of the field, the magnetic flux linked with the coil changes continuously with respect to time. Hence an induced e.m.f. is produced in the coil and a current starts flowing in it. The direction of this induced current is given by Flemming's right hand rule.

(iii) **Expression for Instantaneous e.m.f. Produced :** Figure shows the coil or armature of the a.c. generator consisting of N turns and placed in a uniform magnetic field $\vec{B}$ such that its axis of rotation is perpendicular to the direction of magnetic field. Suppose, initially the plane of the coil is perpendicular to the magnetic flux–lines with arm AB up and CD down. In this position, the number of flux–lines or magnetic flux passing through its plane is maximum.

When the coil is rotated, the magnetic flux passing through it starts changing. Suppose, at any instant during the rotation of the coil, the plane of the coil makes an angle θ with the vertical or normal to the plane of coil makes an angle θ with respect to the direction of magnetic field or component of B perpendicular to the plane of coil is B cos θ.

If ω is the uniform angular speed of the coil, then $\theta = \omega t$.

If A be the area of the coil, then magnetic flux linked with the coil at this instant $\phi = N(B \cos \theta) A = NAB \cos \theta = NAB \cos$ wt. The effect of number of turns N is to increase the magnetic flux N times.

Differentiating both sides with respect to time, we get

$$\frac{d\phi}{dt} = \frac{d}{dt}(NAB \cos \omega t) = -NBA\omega \sin \omega t$$

If e is the induced emf produced at the instant, t then according to Faraday's law of electromagnetic induction

$$e = -\frac{d\phi}{dt} = -(-NBA\omega \sin \omega t) = NBA\omega \sin \omega t$$

It is clear that magnitude of induced e.m.f. in the coil changes continuously with respect to time t. The maximum value of sin w is 1, so maximum value of e is NBAω. If maximum value NBAw is represented by e_0, then $e = e_0 \sin \omega t$

(iv) The induced e.m.f. pro-duced is function of time and therefore as coil passes through different positions with passage of time, e.m.f. produced changes with t as given by above equation. The e.m.f. would be zero for $\omega t = 0$ (when coil is vertical), positive maximum for $\omega t = 90^\circ$ (when coil is horizontal), zero for $\omega t = 180^\circ$ (when coil is vertical), negative maximum for $\omega t = 270^\circ$ (when coil is horizontal) and again zero for $\omega t = 360^\circ$ (when coil is vertical and returns to its initial position).

(v) **Working of Generator :** The working of the a.c. generator is illustrated with the help of five different positions of the armature ABCD at the instants t = 0, T/4, T/2, 3T/4 and T.

Initially, at t = 0, the plane of armature ABCD is vertical with arm AB up and CD down. During the motion of armature between t = 0 to t = T/2, arm AB moves down and CD moves up. According to Flemming's Right Hand Rule, current in the armature will flow in the direction DCBA. On the other hand, during the motion of armature between t = T/2 to t = T, we find just opposite to what we have seen during t = 0 to t = T/2 *i.e.*, arm AB moves up and CD moves down. Therefore, during this interval, the current in the armature will flow in the direction ABCD.

Moreover, whenever the armature passes through vertical position (two times in one rotation), its arm AB and CD move parallel to the field for a moment *i.e.*, rate of change of magnetic flux is zero. Accordingly, the induced emf $e = (d\phi/dt)$ corresponding to vertical position of the armature is also zero [as $(d\phi/dt)$ is zero] *i.e.*, at time t = 0, T/2 and T, induced emf produced is zero.

Whenever the armature flux is fastest or maximum corresponding to horizontal positions. Thus *at time* (t = T/4) and (3T/4), *the induced e.m.f. produce is maximum.*

(vi) Thus, in the first half rotation of the coil, induced e.m.f. in the coil rises from zero to maximum and then again becomes zero while in next half rotation the induced e.m.f. rises from zero to a maximum in opposite direction and again becomes zero. This process is repeated again and again. When a circuit is connected to this coil, then the induced e.m.f. establishes a voltage across the terminals of the circuit which also changes both in magnitude and direction as the induced e.m.f. *The voltage, whose magnitude changes with time and direction, reverses periodically, is known as alternating voltage.* Just like induced e.m.f., the alternating voltage may also be expressed as

$$V = V_0 \sin \omega t \qquad ...(2)$$

where V_0 is the maximum or peak value of the voltage.

POWER OF AN A.C. CIRCUIT

(i) The rate at which electrical energy is dissipated in a circuit is called as its electric power. It is given by the product of voltage and current. If voltage is measured in volts and current in amp, then power will be in watts. In A.C. circuits, power also depends upon the phase difference ϕ between voltage and current.

(ii) In A.C. circuits, because both the voltage and current change continuously with respect to time, hence power cannot be calculated as such. For an A.C. circuit average power over one cycle is calculated after defining the instantaneous power of the circuit. *The instantaneous power of an A.C., circuit is given by the product of instantaneous voltage and the instantaneously current in it.*

(iii) Suppose the instantaneous voltage and current in an A.C. circuit are given by the following equations

$V = V_0 \sin \omega t$ and $I = I_0 \sin (\omega t - \phi)$

where ϕ is the phase angle through which voltage leads the current. V_0 and I_0 in above equations represent the peak values of voltage and current respectively.

If we assume that the values of V and I in the A.C. circuit remain constant for a small time dt, then small amount of energy consumed is given by

$dW = VI\,dt = V_0 \sin \omega t\, I_0 \sin (\omega t - \phi)\, dt = V_0 I_0 \sin \omega t\, [\sin \omega t \cos \phi \;- \cos \omega t \sin \phi]\, dt$

$= V_0 I_0 [\sin^2 \omega t \cos \phi - \sin \omega t \cos \omega t \sin \phi]\, dt$

Now, $\cos 2\omega t = 1 - 2 \sin^2 \omega t$

or $\sin^2 \omega t = (1 - \cos 2\omega t)/2 \qquad ...(1)$

Also, $\sin 2\omega t = 2 \sin \omega t \cos \omega t$

or $\sin \omega t \cos \omega t = (\sin 2\omega t/2)$

$$\text{Hence } dW = V_0 I_0 \left[\left(\frac{1-\cos 2\omega t}{2}\right)\cos\phi - \left(\frac{\sin 2\omega t}{2}\right)\sin\phi\right]dt$$

$$= \frac{V_0 I_0}{2}[\cos\phi - \cos 2\omega t \cos\phi - \sin 2\omega t \sin\phi]\, dt$$

$$= \frac{V_0 I_0}{2}\left[\cos\phi \int_0^T dt - \cos\phi \int_0^T \cos 2\omega t\, dt - \sin\phi\right.$$

$$W = \frac{V_0 I_0}{2}\left[\cos\phi |t|_0^T - \frac{\cos\phi}{2\omega}|\sin 2\omega t|_0^T + \frac{\sin\phi}{2w}|\cos 2\omega t|_0^T\right]$$

$$= \frac{V_0 I_0}{2}\left[\cos\phi(T) - \frac{\cos\phi}{2\omega}\times 0 + \frac{\sin\phi}{2\omega}\times 0\right] = \frac{V_0 I_0 T}{2}\cos\phi$$

Hence, the average power of the A.C. circuit is

$$P_{av} = \frac{W}{T} = \frac{V_0 I_0 T}{2}\cos \times \frac{1}{T} = \frac{V_0 I_0}{2}\cos\phi$$

$$\text{or } P_{av} = \frac{V_0}{\sqrt{2}} \times \frac{I_0}{\sqrt{2}} \times \cos\phi = V_{rms} I_{rms} \cos\phi \qquad \text{...(2)}$$

(iv) In above equations, *$\cos\phi$ is known as POWER FACTOR of the circuit.* Its value depends upon nature of the circuit. It may be noted here that ϕ is the phase angle by which voltage leads (or lags) the current in an A.C. circuit.

(v) **Different Special Cases**

(a) **A.C. Circuit having R only** – We have read that in this circuit voltage and current are in same phase *i.e.*, $\phi = 0$ and power factor $\cos\phi = 1$

Hence $P_{av} = V_{rms} I_{rms} = V^2_{rms}/R$ ...(3)

Thus in an alternating current circuit having ohmic resistance only, the power is equal to the product of rms voltage and rms current e.g. If a 22 ohm resistance is connected to a supply of 220 virtual volt (or 220 volt rms), then the current in the resistance will be 10 virtual ampere and average power of the circuit will be 2200 watt.

(b) **A.C. circuit having L only :** In this circuit we have seen that $\phi = (\pi/2)$ and power factor $\cos\phi = 0$. Hence

$P_{av} = V_{rms} I_{rms} \cos(\pi/2) = 0$ ...(4)

(c) **A.C. circuit having C only :** In this circuit also;

ϕ = (p/2) and power factor cos ϕ = 0

Hence $P_{av} = V_{rms} I_{rms} \cos(\pi/2) = 0$ ...(5)

(d) **A.C. circuit having L and R :** For such a circuit

$\tan\phi = \frac{\omega L}{R}$ and power factor $\cos\phi = \frac{R}{\sqrt{[R^2 + (\omega L)^2]}}$

$$\therefore P_{av} = V_{rms} I_{rms} \frac{R}{\sqrt{[R^2 + (\omega L)^2]}}$$

$$= V_{rms} \times \frac{V_{rms}}{\sqrt{[R^2 + (\omega L)^2]}} \times \frac{R}{\sqrt{[R^2 + (\omega L)^2]}}$$

$$\text{or } P_{av} = \frac{V_{rms}^2 R}{[R^2 + (\omega L)^2]} \quad ...(6)$$

(e) **A.C. circuit having C and R :** For such a circuit,

$\tan\phi = \frac{1/\omega C}{R}$ and power factor

$$\cos\phi = \frac{R}{\sqrt{[R^2 + (1/\omega^2 C^2)]}}$$

$$\therefore P_{av} = V_{rms} I_{rms} \frac{R}{\sqrt{[R^2 + 1/\omega^2 C^2]}}$$

$$= V_{rms} \times \frac{V_{rms}}{\sqrt{[R^2 + 1/\omega^2 C^2]}} \times \frac{R}{\sqrt{[R^2 + 1/\omega^2 C^2]}}$$

$$\text{or } P_{av} = \frac{V_{rms}^2 R}{[R^2 + (1/\omega C)^2]} \quad ...(7)$$

(f) **A.C. circuit having L, C and R :** For such a circuit,

$\tan\phi = \frac{\omega L - (1/\omega C)}{R}$ and power factor

$$\cos\phi = \frac{R}{\sqrt{[R^2 + (\omega L - 1/\omega C)^2]}}$$

Hence, $P_{av} = V_{rms} I_{rms}$

$$\frac{R}{\sqrt{\left[R^2+(\omega L-1/\omega C)^2\right]}}$$

or $P_{av} = V_{rms}$

$$\frac{V_{rms}}{\sqrt{\left[R^2+(\omega L-1/\omega C)^2\right]}} \times \frac{R}{\sqrt{\left[R^2+(\omega L-1/\omega C)^2\right]}}$$

$$\text{or } P_{av} = \frac{V_{rms}^2 R}{\left[R^2+(\omega L-1/\omega C)^2\right]} \qquad ...(8)$$

WATTLESS CURRENT

(i) If the A.C. circuit consists of either inductance only or capacitance only (ohmic resistance is zero), then the phase difference between the current and voltage is 90° *i.e.,* f = 90°.

Then the average power in the circuit is

$P_{av} = V_{rms} \times I_{rms} \times \cos 90° = 0$

Thus, if the ohmic resistance in an alternating–current circuit is zero, then though current flows in the circuit, yet the average power remains zero i.e., there is no energy dissipation in the circuit. The current in such a circuit is known as *WATTLESS CURRENT.* In actual practice, however, a wattless current in never possible because ohmic resistance can not ve zero in a circuit.

(ii) Let us take the example of an A.C. circuit having inductance only. Curves for instantaneous current I, voltage V and power P are drawn in figure for this circuit. It is clear from figure that *Power Curve P is SYMMETRICAL about the time axis* which shows that for one complete cycle the average power P_{av} is zero. Actually, in one fourth cycle the inductor draws energy from the A.C. source and stores it in the form of magnetic field. In the next one fourth cycle, the magnetic field of the inductor becomes zero *i.e.,* the inductor returns the energy to the source. *Thus inductor does not draw any net energy from the source.*

CHOKE COIL

(i) In a direct current circuit, the current is decreased by introducing a suitable resistance by means of a rheostat. When we do so, the electrical energy will be lost (a loss of I^2R per second) in the form of heat energy in the resistance. The current in an alternating current circuit is, however, reduced by means of an inductance in place of resistance. If a resistance is used, there will be wastage of energy in this case also. But when inductance is used, no electrical energy will be wasted. The reason is that the alternating voltage leads the current by a phase angle ($\pi/2$) and therefore average power consumed will be

$$P_{av} = V_{rms} \times I_{rms} \times \cos(\pi/2) = 0 \quad ...(1)$$

(ii) However, in practice inductance coil also possesses a small resistance *i.e.,* a practical inductance coil may be treated as a series combination of inductance L and a small resistance R and it will consume a small average power given by

$$P_{av} = V_{rms} \times I_{rms} \times \frac{R}{\sqrt{\left[R^2 + (\omega L)^2\right]}} \quad ...(2)$$

where $\frac{R}{\sqrt{\left[R^2 + (\omega L)^2\right]}} = \cos\phi$, is the power factor for practical inductance coil.

Since the ohmic resistance R of the coil is nearly zero and its inductance L is very high, so $\cos\phi = 0$ (approx.). Thus according to eqn. (2), the average power consumed in practical inductance coil is also nearly zero.

Hence a device which can regulate current in an A.C. Circuit without much loss of energy, is called as CHOKE COIL.

TRANSFORMER

(i) The transformer is a device which is used to change the voltage of alternating current. As such they are of two types. When it changes the weak alternating current at high voltage into strong alternating current at low voltage, then it is known *Step down transformer*. On the other hand, if it changes strong alternating current at low voltage into weak alternating current at high voltage then it is called as *Step up transformer i.e.,* a transformer is called as step up or step down type depending upon it increases or decreases the voltage of alternating current respectively.

(ii) **Principle :** It works on the principle of Mutual Induction *i.e.*, when current flowing through a coil or magnetic flux linked with a coil changes, an induced e.m.f. is produced in the other coil.

(iii) **Construction :** It consists of two coils of copper wire wound separately over a rectangular and laminated soft iron core (the core is made by placing soft–iron strips one above the other. These strips are insulated from each other to reduce the eddy currents and hence the loss of energy in the core). These coils are kept insulated from each other as well as from the iron core. The two coils are known as primary coil P and secondary coil S. The a.c. source is connected across the primary coil while transformed voltage or transformer output is obtained across the secondary coil S.

(iv) **Theory :** When alternating current flows through the primary coil, then a magnetic field is produced inside the coil which also varies both in magnitude and direction as per nature of input A.C. source. Due to this, core is first magnetised in one direction and then in opposite in each cycle of current. Hence the electrical energy of the primary coil is transformed into the magnetic energy of the core. Since the secondary coil is also wrapped on the same core and the core is made of soft iron, hence the magnetic flux passing through it also changes continuously due to repeated magnetisation and demagnetisation of the core. As there occurs to leakage of magnetic flux in the process of transformation hence as a result of this, an alternating e.m.f. of the same frequency is induced in the secondary coil. Thus magnetic energy of iron–core is now transformed into the electrical energy of the secondary coil. The magnitude of induced e.m.f. produced in the secondary coil depends upon the ratio of number of turns in two coils.

(v) Let ϕ is the magnetic flux linked with each turn and N_p and N_s are the number of turns in the primary and secondary coils respectively. Because there is no leakage of magnetic flux in the process of transformation, hence same magnetic flux ϕ will be linked with each turn of primary and secondary coils at every instant. Therefore, according to Faraday's law of electromagnetic induction, induced e.m.f.'s produced in primary and secondary coils are given by

$$e_p = -N_p\left(\frac{\Delta\phi}{\Delta t}\right) \text{ and } e_s = -N_s\left(\frac{\Delta\phi}{\Delta t}\right)$$

Hence $\dfrac{e_s}{e_p} = \dfrac{N_s}{N_p}$ ...(1)

(vi) If there is no loss of energy in the primary coil and its resistance is assumed negligible then induced emf produced in the primary coil will be nearly equal to the applied potential difference V_p between its ends. Similarly, because the secondary coil is open hence potential difference across its ends will be equal to emf induced in it *i.e.,* under these ideal conditions.

$$\frac{e_s}{e_p} = \frac{V_s}{V_p} = \frac{N_s}{N_p} = r$$

Here, r is called as *Transformation Ratio*. For step up transformer $r < 1$; but for step down transformer $r < 1$.

(vii) If we assume that there are no energy losses in the process of transformation, then Instantaneous output power = Instantaneous input power

or $V_s.I_s = V_p.I_p$

or $\dfrac{I_p}{I_s} = \dfrac{V_s}{V_p} = \dfrac{N_s}{N_p} = r$

Thus, a step-up transformer increases the voltage by decreasing the current and step–down transformer decreases the voltage by increasing the current. This happens according to law of conservation of energy. In other words, *a transformer simply transforms the voltages and currents but does not generate electricity.*

(viii) In real practice, the energy obtained from the secondary coil is always little bit less than the energy given to primary coil. This is because of the fact that some energy of the primary coil is spent in heating its wire & some is spent in the process of magnetisation and demagnetisation of core.

(ix) *Use of transformer for long distance Transformation of electric energy :*

(a) When electrical energy is to be transmitted to distant place, the resistance of line wires becomes considerable. As a result of this, appreciable amount of electrical energy

(= I^2Rt) is lost as heat in the line wires (where I is the current through line wires and R is the resistance). Moreover, a sufficient fall of potential IR occurs along the line wires. Hence, the voltage received at the distant place will be much smaller than that actually supplied by the transmitting station. In addition to this, few more difficulties may arise, if transmission of electrical energy is done at low voltage *i.e.,* at 220 volts.

(b) However, if the transmission of electrical energy is done at high voltage (say 22000 volts) the difficulties faced in the transmission at low voltage can be minimised. To understand the problem, let us take a concrete example. Suppose 11000 watt of electric power is first transmitted at 220 volts and then at 22000 volts.

(c) When transmission of electrical energy is done at 220 volts, then a current of 11000/220 *i.e.,* 50 amp flows through the line wires. If R is the resistance of line wires, energy equal to $(50)^2R$ *i.e.,* 2500 R joules be dissipated per second as heat energy. On the other hand, when transmission is done at 2200 volts, a current of 11000/22000 *i.e.,* 0.5 amp only flows through the wires. In this case, the electric energy dissipated per second as heat will be $(0.5)^2R$ *i.e.,* 0.25R joule per second only.

Therefore we conclude that *if transmission of electric energy is done at high voltage, loss of electric energy as heat energy is very much reduced.*

(d) Further, for the transmission of electric power of 11000 watt at 220 volts, the line wires must have a current capacity of 50 amp and for the transmission of same electric power at 22000 volts, the current capacity of line wires has to be only 0.5 amp. Therefore, for the transmission electric power at low voltage, thick line wires have to be used which will cost more due to their higher weight. Side by side, stronger poles would also be required in order to support thicker, line wire. The conclusion of all this discussion is that *cost of transmission at low voltage will be quite high i.e., from economical point of view also it necessary that transmission should be done at high voltage.*

(e) Thus current from power houses is transmitted through the line wires after stepping up to a very high voltage by means of a step–up transformer. Now, at the place, where the supply is required, it is again stepped down by a step–down transformer.

RESONANT CIRCUITS

(A) Series Resonant Circuit

(i) Consider an alternating current circuit having inductance L, capacitance C and resistance R joined in series. We have read that impedance of this circuit is given by

$$Z = \sqrt{\left[R^2 + (X_L - X_C)^2\right]} = \sqrt{\left[R^2 + (\omega L - 1/\omega C)^2\right]}$$

If the phase difference between the applied voltage and the resulting current is ϕ, then

$$\tan\phi = \frac{X_L - X_C}{R} = \frac{\omega L - 1/\omega C}{R}$$

When inductive reactance X_L and the capacitive reactance X_C of the circuit become equal, then $\phi = 0$ *i.e.,* voltage and the current comes in same phase. Under this condition, impedance Z of the circuit is minimum (= R) and the current I becomes maximum (= V/R). This is called as the condition of resonance *i.e.,* for resonance we have

$$X_L = X_C \text{ or } \omega L = \frac{1}{\omega C}$$

$$\text{or } \omega = \frac{1}{\sqrt{[LC]}}$$

$$\text{or } \phi = \frac{\omega}{2\pi} = \frac{1}{2\pi\sqrt{[LC]}}$$

where f is the frequency of applied voltage. For resonance, f must be equal to $1/\left(2\pi\sqrt{[LC]}\right)$, but because $1/2\pi\sqrt{[LC]}$ represents the natural frequency of circuit hence in a *series resonance circuit frequency of the applied voltage is equal to the natural frequency of the circuit.*

(ii) The important feature of a series resonant circuit is that *voltage available across the inductance L and across the capacitance*

C may be much more than the applied voltage. This is the reason that it is also called as Voltage Resonant Circuit.

The understand this fact, let us take a concrete example. Suppose in series L–C–R circuit, R = 5 ohms,

$X_L = X_C = 20$ ohms and $V_{rms} = 100$ volt

Then current flowing through the circuit is

$I = \frac{V}{Z} = \frac{V}{R} = \frac{100}{5} = 20$ amp [In the state of resonance, Z = R]

∴ Potential difference across the resistance $R = V_R = IR = 20 \times 5 = 100$ volt

Potential difference across the inductance

$L = V_L = IX_L = 20 \times 20 = 400$ volt

and potential difference across the capacitance $C = V_C = IX_C = 20 \times 20 = 400$ volt.

Thus V_L and V_C are much more than the applied voltage V *i.e., Voltage Amplification is obtained in a series resonant circuit.*

(iii) *The main application of series resonant circuit is our radio receiver.* Our radio consists of a series inductive capacitive circuit (which is actually a series L–C–R circuit because both the inductor and capacitor has got some ohmic resistance) which is connected to the Antenna. Electromagnetic waves of different frequencies coming from different radio by turning its knob, then we adjust the capacitance of this circuit (the said circuit consists of a capacitor of adjustable capacitance) in such a way that the circuit is in resonance with the frequency of a particular radio station. In this situation, the current corresponding to that frequency is very much increased (while other currents corresponding to other frequencies become quite weak) and we are able to listen the programme transmitted by that station only.

(B) Parallel Resonant Circuit

(i) Consider an alternating current circuit having inductance L and capacitance L and capacitance C connected in parallel. Let the resistance R be same in both the branches. The inductive reactance and the capacitive reactance of the two branches are $X_L = \omega L$ and $X_C = (1/\omega C)$ respectively. Suppose the main current of the circuit is I and currents through the branches having inductance and capacitance are I_1 and I_2 respectively.

(ii) Because the two branches in the circuit are joined in parallel, hence potential difference across each will be same and will be equal to applied potential difference V.

As impedances, of two branches are

$$\sqrt{[R^2 + X_L^2]} \text{ and } \sqrt{[R^2 + X_C^2]}$$

respectively hence currents I_1 and I_2 flowing through them are given by

$$I_1 = \frac{V}{\sqrt{[R^2 + X_L^2]}} \text{ and } I_2 = \frac{V}{\sqrt{[R^2 + X_C^2]}}$$

If ϕ_1 is the phase difference between I_1 and V, and ϕ_2 that between I_2 and V, then we can write

$$\tan \phi_1 = \frac{X_L}{R} \text{ and } \tan \phi_2 = \frac{X_C}{R}$$

The current I_1 lags behind the potential difference V in phase by phase angle ϕ_1 while I_2 leads the potential difference by phase angle ϕ_2. The appropriate phasor diagram is shown in figure.

(iii) When values of inductance L, capacitance C and frequency w are such that $X_L = X_C$ *then* $I_1 = I_2$. Moreover, if ohmic resistance R is very small as compared to X_L and X_C, then $\phi_1 = \phi_2 \cong 90^o$ (Fig.). It is clear from figure that *the branch currents I_1 and I_2 are much larger than the main current I. Hence in this circuit, current amplification is obtained, and the circuit is called as parallel resonant circuit. This circuit is also called as Current Resonant circuit.*

(iv) *A peculiar feature of this circuit is that smaller is the ohmic resistance R, more closer will be ϕ_1 and ϕ_2 to 90°, and smaller will be the main current I.*

(v) Here also, the condition of resonance is the same as found in series resonant circuit *i.e.,*

$$X_L = X_C \text{ or } \omega L = \frac{1}{\omega C}$$

$$\text{or } \omega = \frac{1}{\sqrt{LC}}$$

$$\text{or } \phi = \frac{1}{2\pi\sqrt{LC}} \text{ (when R is negligible)}$$

(vi) It is also very clear that this circuit does not allow that alternating current to pass through whose frequency is equal to its natural frequency. Hence by using a parallel resonant circuit, out of many currents, the current of one particular frequency may stopped [a observation, which is just opposite to that found in series resonant circuit. Because in series resonant circuit, the circuit allows the current of one particular frequency (= Natural frequency of the circuit)] *i.e.*, the circuit may be used as a *Filter Circuit*. This the reason that this circuit is used in Radio Transmitter.

SOME FORMULAE

1. **Faraday's laws :**

   (i) The induced e.m.f. is directly proportional to the rate of change of flux, associated with circuit,

   i.e., $I. \propto \frac{\Delta\phi}{\Delta t}$ or $e = -K\frac{\Delta\phi}{\Delta t}$.

   In C.G.S., $e = n\frac{\Delta\phi}{\Delta t}$.

   Units : n = number of turns, $\Delta\phi \rightarrow$ maxwell, e $\rightarrow$ ab volt, $\Delta t \rightarrow$ sec. **In S.I.** $\Delta\phi \rightarrow$ weber, e $\rightarrow$ volt, $\Delta t \rightarrow$ sec.

   Relation : 1 weber = 10^8 maxwell.

   (ii) Induced current $(I) = \frac{e}{R} = \frac{1}{R}.\frac{\Delta\phi}{\Delta t}$.

   (iii) Amount of charge that will flow $(q) = I \times \Delta t = \frac{\Delta\phi}{R}$.

2. **Combined form of both Faraday's & Lenz's law :**

   $e = -\frac{\Delta\phi}{\Delta t}$ (one turn), $e = -n\frac{\Delta\phi}{\Delta t}$ (n turns).

3. **Self induction :** $e_1 = -L\frac{\Delta i}{\Delta t}$.

   Where $\frac{\Delta i}{\Delta t}$ = rate of change of current, e_1 = induced e.m.f. & L = co-efficient of self inductance.

   S.I. unit of L is ohm × sec = henry = 10^9 e.m.u. (ab henry).

4. **Mutual inductance :** $e_2 = -M\frac{\Delta i}{\Delta t}$.

Where M is mutual inductance. **S.I. unit** → henry.

5. If a conductor moves in a magnetic field e.m.f. induces across it's ends

   (a) For translatory motion $e = Bvl \sin \theta$

   (b) For Rotatory motion $e = 1/2\ Bl^2 \omega$

   where l = length of conductor, w = Angular velocity of conduct, v = linear velocity

   θ = linear velocity, θ = Angle made by conductor with the field.

Transformer

1. **For an Ideal Transformer :**

   $$V \propto N \text{ or } I \propto \frac{I}{N} \text{ or } \frac{V_2}{V_1} = \frac{I_1}{I_2} = \frac{N_2}{N_1}$$

   or I_1V_1 (input power) = I_2V_2 (output power), where N_2 and N_1 = number of turns in secondary and primary coils, V_2 (output) and V_1 (input) are voltage across secondary & primary and I_2 (output) and I_1 (inp'ut) are currents across secondary & primary.

2. $\frac{N_2}{N_1}$ = Transformation Ratio = k.

   $\frac{N_2}{N_1}$ = Transformation Ratio = k

   for step up transformer k > 1, for step down transformer k < 1.

3. $\text{Efficiency} = \frac{\text{output wattage}}{\text{input wattage}}$.

   $$\text{Percentage efficiency} = \frac{\text{output wattage}}{\text{input wattage}} \times 100.$$

Alternating (Sinusoidal) Current

1. Induced e.m.f. (e) $= -\frac{d\phi}{dt}, = -\frac{d}{dt}(nAB \cos \omega t)$, nABw sin

   $\omega t., = e_0 \sin \omega t.$

   Where $e_0 = nAB\omega$

   Induced current, $i = \frac{e}{R} = \frac{-N}{R}, \frac{d\phi}{dt}$

Where N = Number of turn.

Induced charge dq = i dt

$$= -\frac{N}{R}d\phi, = -\frac{N}{R}(\phi_1 - \phi_2)$$

Obviously charge induced is independent of time. Induced power P = ei

$$= e \times \frac{e}{R}, = \frac{e^2}{R}, = i^2R = l^2B^2v^2/R.$$

2. Induced e.m.f. of (e) of a coil having self-inductance (L)

$$e = -\frac{d\phi}{dt}, = L\frac{dI}{dt}.$$

For small plane circular coil

$$L = \frac{\phi}{I} = \frac{BAN}{I} = \frac{\mu NINA}{2rI}$$

$$= \frac{\mu N^2 \pi r}{2}, \quad \frac{\mu_0 \mu_r N^2 \pi r}{2}$$

Where N = number of turn, r = Radius of coil, μ_0 = Absolute permeability of the medium, μ_r = Relative permeability of the medium.

Also = $\mu_0\, \mu_r\, \pi\, r^2\, n^2 l$, = $\mu_0\, \mu_r\, n^2 V$

Where $n = \frac{N}{l}$ = Number of turn in unit length of coil, V = Volume of coil.

Special case I. When two coils are connected in series having inductance L_1 and L_2, then $L = L_1 + L_2$.

Here we are taking mutual inductance between them (M) = 0.

If M ≠ 0, then $L = L_1 + L_2 + 2M$

When current is flowing in one direction. If M ≠ 0, then $L = L_1 + L_2 - 2M$

When current is flowing in opposite direction.

Case II. When two coils are connected in parallel.

Then

$$L = \frac{L_1L_2}{L_1 + L_2} \text{ or } \frac{1}{L} = \frac{1}{L_1} + \frac{1}{L_2}. \text{ Here } M = 0.$$

If $M \neq 0$, then $\frac{1}{L} = \frac{1}{(L_1 + M)} + \frac{1}{(L_2 + M)}$, So,

$$L = \frac{L_1 L_2 \pm M^2}{L_1 + L_2 \pm 2M}.$$

Case III. Self-inductance of two co-axial cylinder having radii r_1 and r_2.

$$L = \frac{\mu_0}{2\pi r} \log_e \frac{r_2}{r_1}, \quad L = \frac{2.303}{2\pi r} \mu_0 \log_{10} \frac{r_2}{r_1}.$$

3. Mutual inductance (M), $M = \mu_0 n_p N_s A = \frac{\mu_0 N_p N_s A}{l_p}$.

Where N_p = Total number of turn in primary coil,

N_s = Total number of turn in secondary coil,

n_p = Number of turn in unit length of primary coil,

Special case

(f) Mutual inductance between two co-centric rings, having radii of r_s and r_p

$$M = \frac{\pi \mu_0 N_p N_s r_s^2}{2r_p}.$$

Where r_p = radius of primary coil, r_s = radius of secondary coil.

(II) For two paired coil $M = K\sqrt{L_1 L_2}$.

Where K is pair constant of two coil $0 \leq K \leq 1$ or $K = \frac{M}{\sqrt{L_1 L_2}}$.

(III) M in the form of magnetic potential or work

$$M = \frac{2 U_B}{I_0^2} = \frac{2 W_m}{I_0^2}.$$

Where n = number or turns of the coil, A = area of the coil, B = magnetic field induction, w angular velocity of coil = $2\pi f$ and f (frequency) = number of revolutions/sec.

In S.I. : $A \rightarrow m^2$, $B \rightarrow$ weber/m^2, $\omega \rightarrow$ rad/sec & $e \rightarrow$ volt.

4. $e = e_0 \sin \omega t$ & $I = I_0$ win ωt. They are called sinusoidal or periodic or alternating voltage and current.

Where e and I = instantaneous voltage and current, e_0 and I_0 = maximum or peak voltage and current ωt = phase angle, ω

= angular velocity or angular frequency and $\frac{\omega}{2\pi}$ = frequency $(f) = \frac{1}{T}$.

5. *Average voltage and current* : Average value over half cycle = $\frac{2}{\pi}$ × maximum or peak value.

   Hence $I_{av} = \frac{2}{\pi} \times I_0$ & $e_{av} = \frac{2}{\pi} \times e_0$.

6. *R.M.S. or Virtual or Effective value*

   $= \frac{\text{peak value}}{\sqrt{2}}$, *i.e.*, $I_{r.m.s.} = \frac{I_0}{\sqrt{2}}$ and $e_{r.m.s.} = \frac{e_0}{\sqrt{2}}$.

7. *Form factor* $= \frac{\text{R.M.S. value}}{\text{average value}}$

   = 1.1 and is constant.

8. *Phase relations in A.C. circuits* :

   (i) In case of circuit containing only resistance

$$i = \frac{E}{R}.$$

Where E = e.m.f. and R = resistance.

(ii) Current with resistance and inductance (L)

$$i = \frac{E}{\sqrt{R^2 + (L\omega)^2}} = \frac{E}{Z}.$$

Where Z is called impedance and is the effective resistance of an A.C. circuit and $L\omega$ is called inductive reactance.

(iii) Circuit with R and C :

$$i = \frac{E}{\sqrt{R^2 + \left(\frac{1}{C\omega}\right)^2}}.$$

Where $\frac{1}{C\omega}$ = capacitive reactance and impedance

$$(Z) = \sqrt{R^2 + \left(\frac{1}{C\omega}\right)^2}.$$

(iv) Circuit with R, L and C:

$$i = \frac{E}{\sqrt{R^2 + \left(L\omega - \frac{1}{C\omega}\right)^2}}.$$

$$\text{Impedance } Z = \sqrt{R^2 + \left(L\omega - \frac{1}{C\omega}\right)^2}.$$

9. *Condition for resonance :* Frequency (f) $= \frac{1}{2\pi} \cdot \sqrt{\frac{1}{LC}}$.

In this case, $i = \frac{E}{R}$ since $\left(L\omega - \frac{1}{C\omega}\right)^2 = 0$.

Decay time : $t = \frac{L}{R}$, $I = I_0 \left[1 - e^{-\frac{R}{L}t}\right]$,

$$I = I_0 \left[1 - e^{-1}\right] = I_0 \left[1 - \frac{1}{e}\right].$$

$I = 0.63\ I_0$

Where L = Self-inductance of a coil, R = Resistance of coil

FARADAY'S EXPERIMENT ON ELECTROMAGNETISM

With the realisation that electric currents produce magnetic field, Faraday immediately began looking for a reverse effect and suggested that, somehow or other, magnets might also produce electric field.

Faraday's First Experiment

Fig. 1.7 shows a close circuit containing galvanometer. As there is no source of emf so there is no deflection in galvanometer.

Now suppose we move a bar magnet towards the wire keeping the wire stationary, and say north pole facing the wire, the galvanometer needle is deflected indicating that a current has been set in the circuit. However, *the deflection is observed only when the magnet is in motion.* If the magnet is moved away from the wire, the galvanometer needle is deflected in the opposite direction to that when the magnet moves towards the wire indicating that the current set up in the circuit is in opposite direction. It is also observed that if we move the magnet towards the wire, with its south-pole facing the wire, the deflection is in opposite direction, again indicating that the current set up is now in reverse direction to that when the north pole faces the wire.

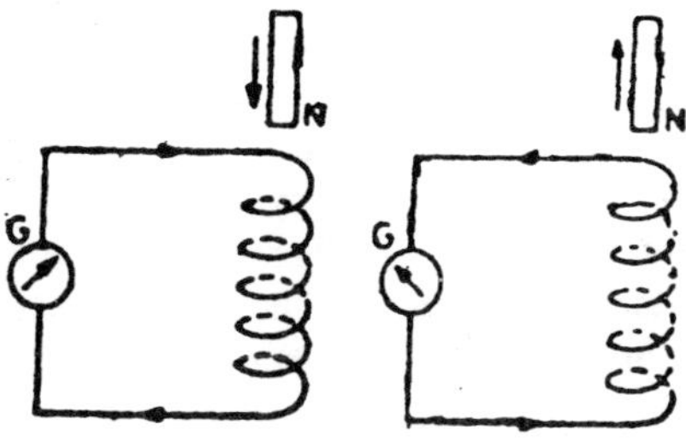

Fig. 1.7

There is also a deflection in the galvanometer when the magnet is held stationary and wire is moved towards or away from magnet. *Faraday discovered that moving the magnet towards the wire one way has the same effect as moving the wire towards the magnet the other way. Thus what actually matters is the relative motion*-between the circuit and the magnet, it is immaterial which of them is moved.

Faraday's Second Experiment

The arrangement is shown in Fig. 1.8. The magnet is replaced by a coil carrying current, we expect to observe the same effect because the coil carrying current produces magnetic field. P is a primary coil carrying current, S is a secondary coil and a galvanometer is connected in series with secondary. The motion of the either of the coil shows a deflection in the galvanometer. He also observed that galvanometer shows a sudden deflection in one direction when current was started in primary and in the opposite direction when current was stopped.

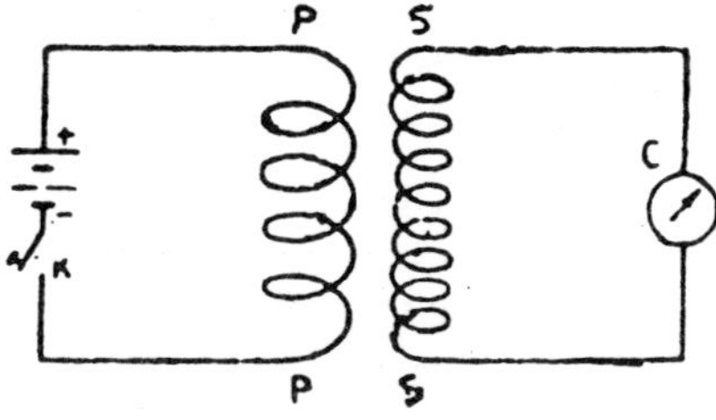

Fig. 1.8

EXPLANATION OF FARADAY'S EXPERIMENTAL OBSERVATIONS

(i) Faraday himself explained the phenomenon of electromagnetic induction on the basis of magnetic flux. In the absence of relative motion between the magnet and the loop of wire, a definite amount of magnetic flux is linked with the later. However, when magnet starts coming closer to the loop due to relative motion, the magnetic flux linked with the loop also starts increasing. It is because of the fact that concentration of magnetic lines of forces is higher near the poles of the magnet. Hence it follows that magnetic flux linked with the loop of wire starts increasing or decreasing as the loop of wire and magnet begin to come close to each other or begin to get away from each other. *Thus, production of induced emf in the loop due to relative motion between the loop and the magnet is, in fact, due to change taking place in the magnetic flux linked with the loop.*

(ii) When the relative motion between loop and magnet becomes zero, the magnetic flux linked with the loop also stops changing and therefore current becomes zero *i.e.*, current flows through the loop till magnetic flux linked with it changes or till there is a relative motion between magnet and loop.

(iii) If the loop or coil is in open circuit (*i.e.*, resistance is infinite), induced emf will still be produced but current will not flow. This shows that *change in magnetic flux linked with the loop induced the emf but not the current.*

(iv) Higher is the speed of magnet towards or away from the loop, higher will be the rate of change of magnetic flux linked with the loop and higher will be the emf induced or higher is induced current produced in the loop or higher is the deflection in the galvanometer. This shows that *magnitude of induced emf is directly proportional to the rate of change of magnetic flux.*

(v) If the magnet is moved away from the loop, then magnetic flux linked with the loop starts decreasing and direction of emf or current induced in the loop gets reversed. This observation tells us about the sign of induced current or induced emf produced. We shall study it later with the help of Lenz's law.

FARADAY'S CONCLUSION

In both experiments, *Faraday found that electric effects exist only when there is something changing.* There was no steady deflection in the galvanometer, but the needle of galvanometer is suddenly deflected due to induced currents produced in the circuit in three different ways,

1. By moving a magnet near wire, towards or away from the wire.
2. By moving the wire towards or away from the magnet, or
3. By changing a current in the adjacent current.

LENZ'S LAW

So far we have not specified the direction of induced emf or current. The direction of the current and emf is given by Lenz's law. It states that the direction of induced emf (or current) in a close circuit is such that it opposes the very cause that produces it. This law is based on the principle of conservation of energy. To be more specific, this law says that the current induced is in a direction such that it produces a magnetic flux tending to oppose the original change of flux *i.e.*, tending to keep the total flux through the circuit constant.

Thus when the applied flux density $\vec{B}$ is decreasing in magnitude the current induced in the loop, is in such a direction as to produce a field which tends to increase $\vec{B}$. On the other hand, when $\vec{B}$ is increasing, the current induced in the loop is in such a direction as to produce a field which tends to decrease $\vec{B}$. Thus the induced current in the loop is always in such a direction as to produce flux opposing the change in $\vec{B}$. Lenz's law applies only to closed circuit.

Consider the Faraday's experiment as shown in Fig. 1.7. As we push the magnet towards the circuit an induced current is set up in the circuit. According to Lenz's law, the direction of this current should be such as to oppose the motion of the magnet towards the circuit. As we know that a current through a coil produces magnetic field like a magnetic dipole. Therefore the two faces of the coil will behave as two poles of a magnet. Thus if the current produced in our coil is to oppose the motion of the magnet towards the coil. The face of the coil towards the magnet should become north pole, similarly in other cases.

(i) Lenz's Rule is a simple method for determining the direction of induced current in a closed circuit. It is based on law of conservation of energy and explains the appearance of negative sign in Faraday's second law of electromagnetic induction.

(ii) According to Lenz's Rule, *the induced current produced in a closed circuit always flows in such a direction that it always opposes the CAUSE due to which it has been produced itself.*

(iii) Faraday's experiments confirm Lenz's Rule. In these experiments, the induced current in the loop is produced due to motion of magnet towards or away from the loop. We have seen that when north pole of the magnet is brought closer to the loop then direction of current in loop is such that face of loop towards magnet behaves as a north pole *i.e.,* it repels the magnet coming towards itself *i.e.,* it opposes the motion of magnet which is the cause of generation of induced current. Similar is the situation when north pole of magnet is moved away from the loop or south pole of the magnet is brought closer or moved away from the loop *i.e.,* in every case current in the loop flows in such a direction that it opposes the motion of magnet.

(iv) Because in each case induced current opposes the motion of magnet, hence some mechanical work has to be done on the system to move the magnet against this opposing force. According to law of conservation of energy, this work is obtained in the coil in the form of heat energy.

(v) Faster is the motion of magnet, higher will be the rate of doing work and higher will be the rate of production of heat energy in the coil *i.e.,* higher will be the induced current.

(vi) If induced current was not to oppose the motion of magnet, then we could be getting electrical energy continuously without doing any work which is never possible. Hence Lenz's rule is the necessity of law of conservation of energy.

INDUCTION AND RELATIVE MOTION

Faraday discovered that moving the magnet towards the wire one way has the same effect as moving the wire towards the magnet—the other way. But when the magnet is moved, we no longer have any $\vec{\upsilon} \times \vec{B}$ force on the electrons in the wire. This is the new effect that Faraday found. Today we understand it from a relativity argument. Thus what actually matters is the relative motion between wire and magnet. We shall discuss few experiments.

A Conducting Rod Moves Through a Uniform Magnetic Field

Fig. 1.9 shows a thin rod or a wire supposed to be moving at constant velocity v in a direction perpendicular to its length, in the space where a uniform magnetic field $\vec{B}$ exists. Let $\vec{B}$ be produced by a solenoid which is at rest in the frame of reference F with coordinates (x, y, s). There is no electric field in the frame F in the absence of the rod.

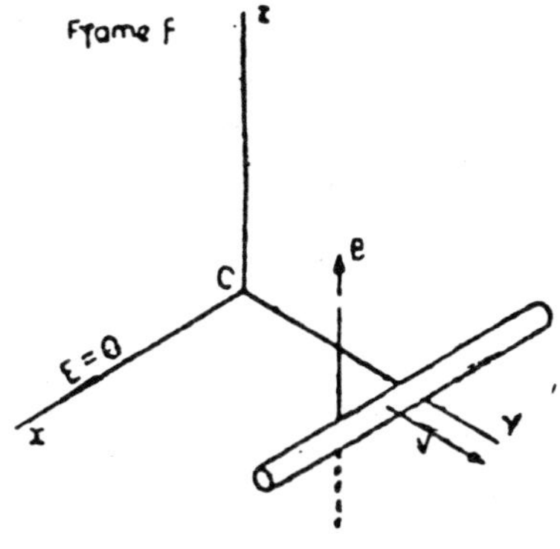

Fig. 1.9

The rod being a conductor, contains charged particles. Those will be in motion when the rod is moved with the same velocity with which the rod is moving. Then a particle of charge q will experience a force,

$$\vec{F} = q(\vec{\upsilon} \times \vec{B})$$

with $\vec{B}$ and $\vec{\upsilon}$ as directed in Fig. 1.10 F is in the positive x-direction if q is positive. In the caw of negatively charged electrons that are in fact the movable carriers in most conductors, the

$\vec{F}$ will be in a direction opposite to that as shown in Fig. 1.10.

The force $\vec{F}$ pushes electrons towards one end of the rod, consequently the other end develops a positive charge. This goes on until these separated charges themselves create an electric field inside the rod. Finally a steady state is reached by moving the rod at constant velocity $\vec{\upsilon}$ when the force $\vec{F}$ is equal to the force due to electric field created inside the rod in the opposite direction. Then the motion of charge relative to the rod ceases, *i.e.*, the field created $\vec{F}$ is given by,

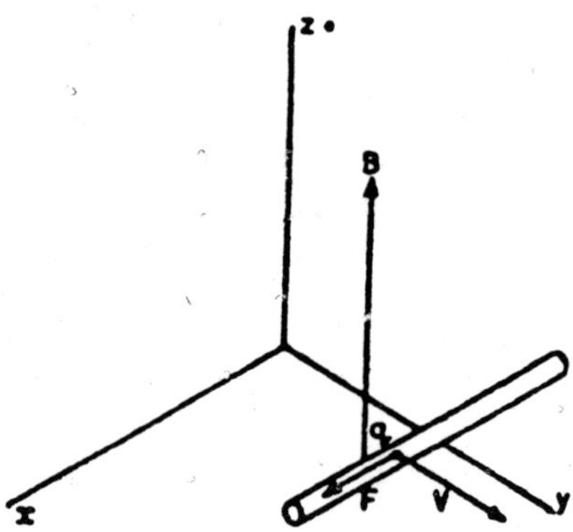

Fig. 1.10

$$q\vec{E} = \vec{F} \qquad ...(1)$$

Thus we can say that the electric field is produced by moving the rod in magnetic field $\vec{B}$ given by,

$$q\vec{E} = -q\,(\vec{\upsilon} \times \vec{B})$$

$$\vec{E} = -\,(\vec{\upsilon} \times \vec{B}) \qquad ...(2)$$

where v is the velocity with which rod is moving in uniform magnetic field B. Thus the charge distribution causes an electric field outside the red.

IMPORTANT POINTS

1. *Inductive reactance (X_L)* is the resistance offered by an inductor and is given by

   $$X_L = \omega L = 2\pi v L$$

   Through a pure inductor, alternating current lags behind the alternating e.m.f. by phase angle of 90°.

2. *Capacitative reactance* (X_C) is the resistance offered by a capacitor and is given by

   $$X_C = \frac{1}{\omega C} = \frac{1}{2\pi v C}$$

   Through a pure capacitor, alternating current leads the alternating e.m.f. by a phase angle of 90°.

   X_L and X_C both are measured in ohm.

3. The total effective resistance of RLC circuit is called *Impedance* (Z) of the circuit. It is given by

   $$Z = \sqrt{R^2 + (X_L - X_C)^2}.$$

   The angle by which alternating voltage leads the alternating current in RLC circuit is given by

   $$\tan \phi = \frac{X_L - X_C}{R}.$$

4. When a condenser of capacity C charged to a certain potential is connected to an inductor L, energy stored in C oscillates between L and C. The frequency of *energy oscillations* is given by

   $$v = \frac{1}{2\pi\sqrt{LC}}$$

   If there is no loss of energy, total energy in L and C at every instant remains constant.

5. A circuit in which inductance L, capacitance C and resistance R are connected in series and the circuit admits *maximum current*, is called series *resonant circuit*. The necessary condition for resonance is $X_L = X_C$,

   which gives $$v = \frac{1}{2\pi\sqrt{LC}}$$

   i.e., frequency of a.c. fed to the circuit becomes equal to natural frequency of energy oscillations in the circuit.

   Under this condition,

   $$Z = R, \; I_0 = \frac{E_0}{Z} = \frac{E_0}{R}$$

   The series resonance circuit is used as an *acceptor* circuit in radio and T.V. receiver sets. The *sharpness of tuning* at *resonance* is measured by *Q factor* or Quality factor of the circuit. It is given by

   $$Q = \frac{1}{R}\sqrt{\frac{L}{C}}.$$

6. *Average power* associated over a complete cycle in a non inductive circuit (containing R, L, C with $X_L = X_C$) or pure resistive circuit is $P = E_v I_v$

On the contrary, average power associated over a complete cycle with a pure inductor or pure capacitor is zero. Hence current through L and through C is said to be *Wattless* or *Idle current.*

In an inductor, energy stored during growth of current is $\frac{1}{2}LI_0^2$. This is spent in maintaining the current during decay.

In a capacitor, energy stored during charging is

$$\frac{1}{2}CV^2 = \frac{1}{2}QV = \frac{1}{2}\frac{Q^2}{C}$$

This is spent during discharging of the condenser. In an LCR circuit, average power associated over a complete cycle is

$P = E_v I_v \cos\phi$,

where ϕ is phase angle between alternating voltage and alternating current in the circuit.

$\cos\phi = P/E_v I_v$ = true power/apparent power, is called *power factor* of the a.c. circuit. The value of power factor varies from 0 to 1.

7. *An a.c. generator/dynamo* produces alternating current energy from mechanical energy of rotation of a coil. It is based on the phenomenon of EMI. The form of e.m.f. induced is $e = e_0 \sin \omega t$, where e_0 = NAB ω, max. e.m.f. induced. Here, N is total number of turns in the coil, A is face area of the coil, B is strength of magnetic field applied and ω is angular velocity of the armature coil.

   The a.c. supply in our country is 220 volt, 50 c/s.

8. *A d.c. generator/dynamo* produces direct current energy from mechanical energy of rotation of a coil. Its principle and working are same as those of a.c. generator. There is only a little change in the design of the generator. *Slip Ring* arrangement used in a.c. generator is replaced by *Split Ring* arrangement in d.c. generator.

9. The negative sign indicates the opposing nature of induced e.m.f. *i.e.,* the current due to induced e.m.f. flows in such a direction, so that it opposite the change in magnetic flux.

10. Self inductance of a plane coil

$L = \frac{\mu_0}{2} \pi N^2 R$ where R is the radius of the circular coil & N is the number of turns.

11. Faraday discovered the phenomenon of E.M.I. and established the following two laws of E.M.I.

    First Law. Whenever the amount of magnetic flux linked with a circuit changes, an e.m.f. is induced in the circuit. This induced e.m.f. lasts so long as the change in magnetic flux continues.

    Second Law. The magnitude of e.m.f. induced in a circuit is directly proportional to the rate of magnetic flux linked with the circuit, *i.e.*,

$$e = \frac{\phi_2 - \phi_1}{t} \quad \text{or} \quad e = \frac{-d\phi}{dt}.$$

12. The direction of induced e.m.f. is given by *Lenz's law.* According to this law, the direction of induced e.m.f. in a circuit is always such as to oppose the change in magnetic flux responsible for it.

    Lenz's law is in accordance with the principle of conservation of energy. Infact, work done in moving the magnet w.r.t. the coil changes into electric energy producing induced current.

    There is also another law for finding the direction of induced current. This is *Fleming's right hand rule.* According to this rule, if we stretch the first finger, the central finger and thumb of our right hand in mutually perpendicular directions, such that first finger points along the magnetic field and thumb points along the direction of motion of the conductor, then central finger would give us the direction of induced current.

13. The coefficient of self inductance of the solenoid is $L = \frac{\mu_0 N^2 A}{l}$ where l is the length of the solenoid.

14. The use of the conducting coopering for the coil in the dead beat galvanometer closely follows the Lenz's law, as the induced current in the ring opposes the relative motion of the coil with respect to the magnetic field and due to which the current is induced.

15. Current flowing in the circuit due to the induced e.m.f. will also be alternating nature. It can be expressed by the following equation $I = I_0 \sin \omega t$.

16. The electric currents which vary for a short finite time, while growing from zero to maximum value or while decaying from maximum value to zero value are called transient currents Helmholtz equation for growth of current in RL circuit is I = I_0 (1 – $\left(e^{\frac{-Rt}{L}}\right)$ and for decay of current in RL circuit is

$$I = I_0 \, e^{-\frac{Rt}{L}}$$

Time constant of LR circuit is $\tau = \frac{L}{R}$.

17. Inductive time constant of the L-R circuit is defined as the time during which current grows to 63.2% of its maximum value of reduces to 36.8% of its maximum value.

18. Charging of a condenser of capacity through a resistance R is given by $q = q_0\,(1 - e^{t/RC})$.

19. The discharging same capacitor through same resistance R is given by $q = q_0\,e^{-t/RC}$.

20. Energy stored in a self inductance coil $U = \frac{L\,I_0^2}{2}$ where U is the potential energy stored in the coil.

21. Magnetic flux $\phi = nBA\cos\theta$ where n is no. of turns in the coil. A is face area of the coil B is magnetic field strength and Q is the angle which normal to the coil makes with the direction of B.

22. Whenever two coils are placed near each other and the current flowing in one of them changed then an emf is induced in the second coil. This phenomenon of electromagnetic induction is called mutual induction.

23. For a solenoid having a primary coil of N_1 turns and a secondary coil of N_2 turns, the coefficient of mutual inductance is given by $M == \frac{\mu_0\,\mu_r\,N_1\,N_2\,A}{l}$.

24. If the current carrying conductor moves with a velocity v at an angle θ with the field direction, then the component of velocity perpendicular to $\vec{B}$ is V sin θ. In this case the induced e.m.f. across the ends is $Bvl\sin\theta$.

25. Induced e.m.f. is developed only when $\vec{v}$, $\vec{B}$, and $\vec{l}$ are mutually perpendicular to each other.

26. Self inductance of a coil is called inertia as it opposes the variation of current.

27. If the coil is closed and the total resistance of the coil is R, then induced current in the coil is

$$i = \frac{e}{R} = \frac{N}{R} - \frac{d\phi}{dt}.$$

28. Induced charge flowed through the circuit in time interval dt is

$$q = idl = \frac{N}{R} d\phi.$$

Induced charge is independent of the time interval of flux change.

29. Electromagnetic induction is the phenomenon of producing an electric current in a coil by changing the amount of magnetic flux linked with the coil the current so produced is called induced current and the corresponding e.m.f. is called induced e.m.f.

30. Faraday's laws of electro magnetic induction

    (a) Whenever magnetic flux linked with a circuit changes, an induced e.m.f. is always produced in it.

    (b) The magnitude of the induced e.m.f. is directly proportional to the rate of change of the magnetic flux $e = -\frac{d\phi}{dt}$.

31. A *d.c. motor* converts direct current energy from a battery into mechanical energy of rotation of a coil. Its working principle is that when a coil carrying current as held in a magnetic field, a torque acts on the coil, which rotates the coil.

    An e.m.f. (E) is induced in the coil on account of rotation of the coil in the magnetic field. This called *back e.m.f.* as it opposes the rotation of the coil. If V is e.m.f. of battery and R is resistance of armature coil, then $I = \frac{V - E}{R}$.

    Efficiency of *d.c. motor* = *output mechanical power/input electrical power. This is equal to back e.m.f./e.m.f. of battery.*

32. *A motor starter* is a safety device, which protects the motor coil from burning out, when the motor is switched on. It is a *variable resistance* connected in series with the coil of d.c. motor.

33. *A choke coil* is the best method of reducing a.c. in a circuit, because it involves no dissipation of power. A choke coil consists of a few turns of thick insulated copper wire wound over an air core/iron core. For reducing high freq. a.c., air cored chokes are used, and for reducing low frequency a.c.; iron cored chokes are used.

34. Time constant of CR circuit is $\tau = RC$.

35. Electromagnetic induction (EMI) is the phenomenon of producing an electric current in a coil by changing the amount of magnetic flux linked with the coil. The current so produced is called *induced current* and the corresponding e.m.f. is called *induced e.m.f.*

36. Magnetic flux $\phi = n\ B\ A \cos\theta$, where n is no. of turns in the coil, A is face area of the coil, B is magnetic field strength and θ is the angle which normal to the coil makes with the direction of B.

37. Induced e.m.f. is temporary. It lasts so long as change in magnetic flux continues. Magnitude of induced e.m.f. $e = \frac{d\phi}{dt}$ = rate of change of magnetic flux. The direction of induced e.m.f. is given by Lenz's law/Fleming's right hand rule. It always opposes the change in magnetic flux responsible for its production.

38. Three methods of inducing e.m.f. are by changing B or by changing A or by changing θ.

39. Currents induced in the body of a conductor when the amount of magnetic flux lined with the conductor changes are called *eddy currents*

$$i = \frac{e}{R} = \frac{-d\phi / dt}{R}$$

where R is resistance of conductor. The direction of eddy currents is given by Lenzel's law/Fleming's right hand rule. Eddy currents have many advantages and disadvantages.

40. *Self Induction* is the property of a coil by virtue of which it opposes any change in the strength of current flowing through it. Coefficient of self inductance of a coil is calculated from

$$L = \phi/I \text{ and } L = \frac{-e}{dI/dt}.$$

value of L depends on geometry of the coil.

Coefficient of self inductance of a solenoid of length l, area of cross section A having N turns on air core is

$$L = \mu_0 N^2 A/l.$$

41. *Mutual Induction* is the property of two coils by virtue of which each coil opposes any change in the strength of current flowing through the other coil. Coefficient of mutual inductance (M) of two coils is calculated from

$$M = \phi/I \text{ and } M = \frac{-e}{dI/dt}$$

For two co-axial solenoids of same length l & same area of cross section A wound on air core, $M = \frac{\mu_0 N_1 N_2 A}{l}$ when N_1 & N_2 are total no. of turns in the two solenoids. S.I. unit of I & M is henry (H).

42. *An a.c. generator/dynamo* produces alternating current energy from mechanical energy of rotation of a coil. It is based on the principle that whenever the amount of magnetic flux linked with a coil changes, an e.m.f. is induced in the coil. The change in magnetic flux is carried out by rotating the coil continuously in a magnetic field.

43. A *d.c. generator/dynamo* produces direct current energy from mechanical energy of rotation of a coil. Its working principle is same as that of a.c. generator.

    The slip ring arrangement of a.c. generator is replaced by *split ring arrangement* or *commutator arrangement.* The current produced is unidirectional. The magnitude of e.m.f. induced in the coil of a.c./d.c. generator is $e = e_0 \sin \omega t$. The direction of induced e.m.f. is given by Fleming's right hand rule.

44. *A transformer* is an appliance which increases the alternating voltages. The former is called a *step up transformer* and the latter is called a *step down transformer.* A transformer is based on the principle of *mutual induction.* We can show that $\frac{E_s}{E_p} = \frac{n_s}{n_p} = K = \frac{I_p}{I_s}$ where K is called transformer ratio,

E_s, E_p are secondary and primary voltages,

I_s, I_p are secondary and primary currents.

Efficiency of a transformer is given by

$$\eta = \frac{\text{output power}}{\text{input power}} = \frac{E_s I_s}{E_p I_p}.$$

45. As is known, e.m.f. is induced in a circuit only when amount of *magnetic flux* linked with the circuit changes. As $\phi = BA\cos\theta$, therefore, three methods of producing induced e.m.f. are (i), by changing B, (ii) by changing A and, (iii) by changing θ. When a conductor of length l moves with a velocity v in a magnetic field of strength B so that magnetic flux linked with the circuit changes, the e.m.f. induced (e) is given by $e = Blv$.

46. An induction coil is used for producing weak intermittent current at high voltages $\approx 10^5$ volt. It is based on the principle of mutual induction.

47. *Units of self inductance.* We have studied 1 henry as the SI unit of self inductance. Another unit of self inductance is *ab henry* or e.m.u. of inductance. The self inductance of a coilis said to be one ab henry, when a current change at the rate of 1 ab ampere per second causes an induced e.m.f. of one ab volt in the coil.

As $$1 \text{ henry} = \frac{1 \text{ volt}}{1 \text{ ampere/sec}}$$

$$= \frac{10^8 \text{ ab volt}}{\frac{1}{10} \text{ ab ampere/sec}}.$$

$\therefore$ 1 henry = 10^9 ab henry.

48. *Eddy currents* are the currents induced in the body of a conductor when the amount of magnetic flux linked with the conductor changes. These currents are also called *Foucault currents.* The magnitude of eddy currents is given by $i = \frac{-e}{R} = \frac{d\phi/dt}{R}$ where R is resistance of the conductor. The direction of eddy currents is given by Lenz's law or Flemming's right hand rule.

Some of the *important applications* of eddy currents are: Electromagnetic damping, induction furnace, electromagnetic damping, induction motor, speedometers and in diathermy *i.e.*, deep heat treatment of parts of human body.

Some of the *undesirable effects* of eddy currents are that they oppose the relative motion, involve loss of energy in the form of heat and reduce the life of electrical devices. To minimise eddy currents, we use *laminated cores*.

49. *Impedance Triangle* is a right angled triangle whose base represents ohmic resistance R, perpendicular represents reactance $(X_L - X_C)$ and hypotenuse represents impedance Z of the RLC circuit. From this triangle, we can calculate Z and ϕ as stated above.

50. *Self induction* is the property of an electrical circuit by virtue of which the circuit opposes any change in the strength of current flowing through it by inducing an e.m.f. in itself.

When a current I flows through a coil, the magnetic flux ϕ linked with the coil is $\phi = LI$, where L is *coefficient of self induction* or *self inductance* of the coil. On differentiating, we get

$$\frac{d\phi}{dt} = I.\frac{dI}{dl} = -e$$

It $dl/dt = I$; $L = -e$. Hence self inductance of a coil is equal to e.m.f. induced in the coil when rate of change of current through the coil is unity. The value of L depends on *geometry* of the coil and is given by $L = \frac{\mu_0 N^2 A}{l}$

where l is length of the coil (solenoid) N is total number of turns of solenoid and A is area of cross section of the solenoid.

The S.I. unit of L is *henry*. Coefficient of self induction of a coil is said to be one henry when a current change at the rate of 1 ampere/sec. in the coil induces an *e.m.f. of one volt in the coil.*

51. *Mutual Induction* is the property of two coils by virtue of which each opposes any change in the strength of current flowing through the other by developing an induced e.m.f.

If ϕ is the amount of magnetic flux linked with one coil when a current I flows through the other coil, then $\phi = MI$, where M is coefficient of mutual induction of two coils.

As $$e = \frac{-d\phi}{dt}$$

$$= \frac{-d}{dt}(MI) = -M\frac{dI}{dt}$$

$\therefore$ If $\frac{dI}{dt} = 1$, then $e = -M \times 1$ or $M = -e$.

Hence coefficient of mutual induction of two coils is equal to the e.m.f. induced in one coil, when rate of change of current through the other coil is unity. The S.I. unit of M is *henry.* Coefficient of mutual inductance of two coils said to be one henry, when a current change at the rate of 1 ampere/sec. in one coil induces an e.m.f. of one volt in the other coil. The value of M depends on *geometry* of two coils, distance between two coils, relative placement of two coils etc.

The coefficient of mutual inductance of two long co-axial solenoids, each of length l, area of cross section A wound on an air core is

$M = \frac{m_0 N_1 N_2 A}{l}$, where N_1, N_2 are total number of turns of the two solenoids.

52. Equivalent coefficient of self inductance of two coils in series.

 If M is the coefficient of mutual inductance between the two coils when they flüx linkage in same sense $e = L_1 + L_2 - 2M$.

53. Coefficient of self inductance of two coils is parallel. $L = \frac{L_1 L_2}{L_1 + L_2}$.

54. Coefficient of coupling between two coils having self inductance L_1 and L_2 and coefficient of mutual inductance M is

$$K = \frac{\pm M}{\sqrt{L_1 L_2}}.$$

55. *Relation between henry and weber*

 From $\phi = L(I)$, $L = \frac{\phi}{I}$

 $\therefore$ 1 henry = 1 weber/ampere.

56. *Self inductance of a plane coil with air core*

$$L = \frac{\mu_0 \pi n^2 r}{2}$$

where n = number of turns in the coil

r = radius of the coil

57. *Reciprocity theorem of mutual inductance,* says that mutual inductance of two solenoids S_1 and S_2 remains unchanged irrespective of the fact that current is passed through S_1 or S_2 *i.e.,*

$$M_{12} = M_{21}.$$

58. *Electromagnetic Induction* (E.M.I.) is the phenomenon of generating an e.m.f. by changing the number of magnetic lines of force associated with a circuit. The e.m.f. so generated is called *induced e.m.f.* and the corresponding current is called *induced current.*

The number of magnetic lines of force crossing a surface is called *magnetic flux* linked with the surface. It is represented by

$$\phi = \vec{B}.\vec{A} = BA\cos\theta$$

where B is strength of magnetic field, A is area of the surface and θ is the angle which *normal to the area* makes with the direction of magnetic field.

The S.I. unit of magnetic flux is weber which is the amount of magnetic flux over an area of 1 m^2 held normal to a uniform magnetic field of one tesla. The c.g.s. unit of ϕ is *maxwell.* 1 weber = 10^8 maxwell.

59. If the circuit is a coil consisting of N turns and all the turns are quite close to each other, then rate of change of magnetic flux linked with every turn will be same, same e.m.f. will be induced in every turn and all the emf's will be added together. Hence induced emf produced in the whole coil will be

$e = -N\left(\frac{d\phi}{dt}\right)$ If ϕ is measured in webers and time t in second e.m.f. e will be in volts.

60. The basic reason for the generation of induced emf in a circuit is the change in magnetic flux linked with it. The magnetic flux can be changed by relative motion between magnetic and closed coil.

61. An a.c. generator/dynamo produces alternating current energy from mechanical energy of rotation of a coil. It is based on the

principle that whenever the amount of magnetic flux linked wi*h a coil changes, an e.m.f. is induced in the coil the change in magnetic flux is carried out by rotating the coil continuously in a magnetic field.

62. A d.c. generator/dynamo produces direct current energy from mechanical energy of rotation of a coil. Its working principle is same as that of a.c. generator.

63. When a coil is rotated in a magnetic field B about an axis, perpendicular to the magnetic field, an induced e.m.f. is produced in the coil which is given by e = NBA ω sin ωt. Where N is the number of turns in the coil, A is the area of the coil and ω the angular speed of the coil.

64. Induced e.m.f. is a periodic function of time and varies both in magnitude and direction as per nature of the sin θ function. Such an e.m.f. is called as alternating e.m.f.

SOLVED EXAMPLES

Example 1:

A solenoid of resistance 50W and inductance 80 henry is connected to a 200 volt battery. How long will it take for the current to reach 50% of its final equilibrium value? Calculate the maximum energy stored.

Solution:

Given that R = 50Ω, L = 80 H, E = 200 volt

$$I = I_0\left(1-e^{-t/\tau_L}\right) \text{ where } \tau_L = \frac{L}{R} \text{ and } I_0 = \frac{E}{R}$$

Let $I = \frac{I_0}{2} = \frac{E}{2R}$ when $t = \tau_0$,

$$\frac{I_0}{2} = I_0\left(1-e^{-\tau_0/\tau_L}\right)$$

or $e^{-\tau_0/\tau_L} = \frac{1}{2}$

or $e^{-\tau_0/\tau_L} = 2$ or $\tau_0 = \tau_L \log_e 2 = (L/R)\log_e 2$

$$\therefore \ \tau_0 = \frac{80}{50} \times 0.693 = 1.109 \text{ sec}$$

$$E_{max} = \frac{1}{2}LI_0^2 = \frac{1}{2}L\left(\frac{E}{R}\right)^2 = \frac{1}{2} \times 80 \times \left(\frac{200}{50}\right)^2 = 640 \text{ J}.$$

Example 2:

A battery of constant emf. 10V and negligible internal resistance is connected across a coil of inductance 0.5 Henry and resistance 2.5 ohms. Calculate

(a) The current flowing in the coil after 0.2 sec.

(b) Time, after closing the circuit, for the current to reach a value 2 amp.

(c) If after attaining the steady state, the battery is shorted, then what will be the time taken for the current to reduce to half its original value in steady state.

Solution:

(i) Current in L–R circuit at any instant t after closing the circuit, is given by

$$i = i_0\,[1 - e^{-(R/L)t}] = \frac{V_0}{R}\,[1 - e^{-(R/L)t}]$$

$$= \frac{10}{2.5}\,[1 - e^{-(25/0.5)\times 0.2}] = 4\,[1 - e^{-1}] = 4 - (4/e) = 2.528 \text{ amp.}$$

(ii) As $i = 2$ amp, and $i_0 = (V_0/R) = 4$ amp, hence $i = i_0/2$

i.e., $\dfrac{i_0}{2} = i_0[1 - e^{-(R/L)t}]$

or $\dfrac{1}{2} = 1 - e^{-(R/L)t}$

or $\left(\dfrac{1}{2}\right) = e^{-(R/L)t}$

or $e^{(R/L)t} = 2$

or $(R/L)t = \log_e 2$

or $\left(\dfrac{2.5}{0.5}\right) t = 2.303 \log_{10} 2$

$\therefore\ t = \dfrac{0.6931}{5} = 0.1386$ second.

(iii) After shorting the battery

Now, $i = i_0\, e^{-(R/L)t}$

or $(i_0/2) = i_0 e^{-(R/L)t}$

or $(R/L)t = \log_e 2$

$$\text{or } t = \frac{\log_e 2}{(R/L)} = \frac{0.6931}{5} = 0.1386 \text{ seconds.}$$

Example 3:

A coil of self inductance 2 henry and resistance 10Ω are in a closed series circuit with an open key and cell of constant 100 volt with negligible resistance. At time $t = 0$, the key is closed. Find

(a) the time constant of the circuit

(b) the maximum steady current in the circuit

(c) the current in the circuit at $t = 1$ second

(d) the energy stored in the magnetic field linked with the coil in the steady state.

Solution:

Given that L = 2H and R = 10 ohms.

(a) time constant of the circuit

$$= \frac{L}{R} = \frac{2H}{10\Omega} = 0.2 \text{ sec.}$$

(b) Maximum steady current

$$i_0 = \frac{V_0}{R} = \frac{100 \text{ volt}}{10\Omega} = 10 \text{ amp.}$$

(c) $i = i_0\,[1 - e^{-(R/L)t}] = 10\,[1 - e^{-(10/2)\times 1}] = 10$

$[1 - e^{-5}] = 9.933$ amp.

(d) Energy stored in the magnetic field, at the steady state

$$= \left(\frac{1}{2}\right)L i_0^2 = \left(\frac{1}{2}\right) \times 2 \times (10)^2 = 100 \text{ J.}$$

Example 4:

A 22 cm long solenoid having total number of turns 1000 consists of a core cross–sectional area 4 m^2. Half portion of the core consists of air and half portion is made of iron (Relative permeability 500). Calculate the self inductance of solenoid.

Solution:

If a solenoid of length l consists of number of turns N, then magnetic field produced at its centre due to a current i flowing in it, is given by $B = \mu N i / l$

$\therefore$ Magnetic flux linked with the solenoid = NBA

i.e., $\phi = N\left(\frac{\mu Ni}{l}\right)A = \frac{\mu N^2 A}{l}i$ ...(1)

According to definition of self inductance we know that :

$\phi = Li$ ...(2)

Comparing (1) and (2), we get

$L = (\mu N^2 A/l)$ ...(3)

Because core of solenoid consists of two magnetic mediums, hence self inductance

$$L = L_1 + L_2 = \frac{\mu_1 N^2 A_1}{l} + \frac{\mu_2 N^2 A_2}{l}$$

For air $\mu_1 = \mu_0$ and for medium of relative permeability $\mu_r = 500$ we have $\mu_2 = \mu_r \mu_0$

$$L = \frac{\mu_0 N^2}{l}[A_1 + \mu_1 A_2] = \frac{(4\pi \times 10^{-7}) \times (1000)}{0.22}$$

$$[2 \times 10^{-4} + 500 \times 2 \times 10^{-4}] = 0.57 \text{ Henry.}$$

Example 5:

An iron core is inserted into a solenoid 0.5 m long with 400 turns per unit length. The area of cross–section of the solenoid is 0.001 m^2.

(a) Find the relative permeability of the core when a current of 5 amp flows through the solenoid and magnetic flux through the cross section of the solenoid is 1.6×10^{-3} weber.

(b) Also calculate the self inductance of the solenoid.

Solution:

(a) If m_r represents the relative permeability of the medium of the core inside the solenoid the magnetic induction at the axis of the solenoid is given by

$B = \mu_r \mu_0 ni$ ...(1)

Hence magnetic flux through the cross section of the solenoid is given by

$\phi = BA = \mu_r \mu_0 niA$

or $1.6 \times 10^{-3} = \mu_r (4\pi \times 10^{-7}) \times 400 \times 5 \times 0.001$

$$\therefore\ \mu_r = \frac{1.6 \times 10^{-3}}{4\pi \times 10^{-7} \times 2} = 637$$

(b) Total number of turns in the solenoid

$$N = n \times l = 400 \times 0.5 = 200$$

$$\therefore\ L = \frac{Nf}{i} = \frac{200 \times 1.6 \times 10^{-3}}{5} = 0.064\ H = 64\ mH.$$

Example 6:

An air ship 50 feet in diameter has a coil of wire of 100 turns wound round it. It is turning in a circle of 1/4 mile radius at 30 mph. Taking H = 0.18 gauss, calculate the maximum e.m.f. generated in the coil.

Solution:

Suppose at any instant the plane of coil makes an angle θ with the direction of H. Then component of B normal to the plane of coil is H sin θ.

Hence magnetic flux linked with the coil

$$\phi = NHA \sin\theta$$

$$\therefore\ \text{Induced emf} = e = -\frac{d\phi}{dt} = -NHA \cos q\ \frac{d\theta}{dt}$$

or $e = -(NHA \cos\theta)\ \omega = -NHA\omega \cos \omega t$

[$d\theta/dt = \omega$ = Angular speed of the coil which is the same as the angular speed of the ship]. The induced emf will be maximum when $\cos \omega t = 1$

$$\therefore\ e_{max} = NHA\ \omega = \frac{100 \times 0.18 \times \pi(25 \times 3 \times 30.48)^2}{10^8} \times \frac{1}{30}\ \text{volt}$$

$$\left[\therefore\ \omega = \frac{v}{r} = \frac{30}{3600 \times (1/4)} = \frac{1}{30}\ \text{radian/sec}\right]$$

$$\therefore\ e_{max} = 1.094 \times 10^{-2}\ \text{volts.}$$

Example 7:

An a.c. voltage source of E = 150 sin 100 t is used to run a device which offers a resistance of 20 Ω and restricts the flow of current in one direction only. Calculate the average and r.m.s. values of current in the circuit.

Solution:

Here, $E = 150 \sin 100\, t$,

$R = 20\ \Omega$

Compare it with $E = E_0 \sin \omega t$.

$E_0 = 150$ volt

$$I_0 = \frac{E_0}{R} = \frac{150}{20} = 7.5 \text{ A}$$

$$I_{av} = \frac{I_0}{\pi} = \frac{7.5}{\pi} = 2.38 \text{ A}$$

$$I_v = \frac{I_0}{\sqrt{2}\,\sqrt{2}} = \frac{7.5}{2} = 3.75 \text{ A.}$$

Example 8:

How much charge will flow through a 200 Ω galvanometer connected to a 400 Ω circular coil of 1000 turns wound on a wooden stick 20 mm in diameter, if a magnetic field B = 0.0113 T parallel to the axis of the stick is decreased suddenly to zero.

Solution:

Here, total resistance of the circuit,

$R = 200 + 400 = 600\ \Omega$

$n = 1000$, $D = 20$ mm,

$r = 10 \text{ mm} = 10^{-2}\text{m}$.

$\therefore$ Area of cross-section $A = \pi r^2$

$= \pi\, (10^{-2})^2 \text{ m}^2 = 10^{-4}\, \pi \text{ m}^2$

$B_1 = 0.0113$ T, $B_2 = 0$,

$\therefore\ dB = B_2 - B_1 = -\,0.0113$ T

$$\text{Now } e = IR = \frac{dq}{dt} R$$

$$\text{Also, } e = -\frac{n\, d\phi}{dt} = \frac{-n\, A\, d\, B}{dt}$$

$$\therefore\ dq = \frac{-n\, A\, d\, B}{R}$$

$$= \frac{-1000 \times 10^{-4}\pi \times (-0.0113)}{600}$$

$$= +5.9 \times 10^{-6} \text{ C.}$$

Example 9:

The brush potential of a separately excited generator, when it is delivering 5A is 125V. When the generator delivers 15A, the potential difference across the brushes falls to 122 V. What are the induced e.m.f. and the resistance of armature?

Solution:

Here, (i) I = 5A, V = 125V

As $V = E - I\,r \therefore 125 = E - 5r$ ...(1)

(ii) When I = 15A, V = 122 V

$\therefore \quad 122 = E - 15r$ ...(2)

Subtract (2) from (1), 3 = 10r, r = 0.3W

From (1), $125 = E - 5 \times 0.3$

$$E = 125 + 1.5$$

$$= 126.5 \text{ V.}$$

Example 10:

A 200 km long telegraph wire has capacity of 0.14μF/km. If it carries an alternating current of 50 kc/s, what should be the value of an inductance required to be connected in series so that impedance is minimum.

Solution:

Here, Capacity of telegraph wire,

$C = 0.014 \times 200\ \mu F = 2.8 \times 10^{-6}$ F

$v = 50$ kc/s $= 50 \times 10^3$ c/s, L = ?

At resonance, $$v = \frac{1}{2\pi\sqrt{LC}}$$

or $$L = \frac{1}{4\pi^2 v^2 C}$$

$$L = \frac{1}{4 \times \frac{22}{7} \times \frac{22}{7}\left(50 \times 10^3\right)^2 \times 2.8 \times 10^{-6}}$$

$\therefore \quad L = 0.36 \times 10^{-3}$ hertz.

Example 11:

A step-up transformer operates on a 220 volt line and supplies to load, a current of 2A. The ratio of primary and secondary winding is 1 : 25. Calculate secondary voltage, primary current and power output, if efficiency is 80%.

Solution:

Here, $E_P = 220$ V, $I_S = 2$A,

$$\frac{n_P}{N_s} = \frac{1}{25}$$

As $\dfrac{E_S}{E_P} = \dfrac{n_S}{n_P}$

$$\therefore \quad \frac{E_S}{220} = 25, \; E_S = 5500 \text{ V}$$

Output power $= E_S I_S = 5500 \times 2$

$= 11000$ watt.

As $\eta = \dfrac{P_0}{P_i} = \dfrac{E_S I_S}{E_P I_P}$

$$\therefore \quad \frac{80}{100} = \frac{100}{220\, I_P} = \frac{50}{I_P}$$

$$\therefore \quad I_P = \frac{50 \times 5}{3} = 62.5 \text{ A.}$$

Example 12:

An electromagnet has stored 648J of magnetic energy, when a current of 9A exists in its coils. What average e.m.f. is induced if the current is reduced to zero in 0.45 s?

Solution:

Here, $E = 648$ J, $I = 9$A, $e = ?$

$$dI = 9 - 0 = 9A,\ dt = 0.45s$$

From $E = \frac{1}{2} L I^2$

$$648 = \frac{1}{2} \times L\,(9)^2;$$

$$L = \frac{648 \times 2}{9 \times 9} = 16H$$

As $e = \frac{L\,dI}{dt}$ $\therefore$ $e = \frac{16\,(9)}{0.45} = 320\ V$.

Example 13:

A 120 V, 60 Hz power source is connected across an 800 Ω non inductive resistance and an unknown capacitance in series. The voltage drop across the resistor is 102 V. (a) What is the voltage drop across the capacitor? (b) What is the reactance of the capacitor?

Solution:

Here, V = 120 V,

$V_R = 102$ V, $V_C = ?$ $X_C = ?$

As $V_R = IR$ $\therefore$ $I = \frac{V_R}{R} = \frac{102}{800} = 0.128$ A

As $V = \sqrt{V_R^2 + V_C^2}$

$$\therefore\ V_C = \sqrt{V^2 - V_R^2} = \sqrt{120^2 - 102^2} = 63V$$

As $V_C = I\,X_C$

$$\therefore\ X_C = \frac{V_C}{I} = \frac{63}{0.128}\ \ 494\ \Omega.$$

Example 14:

A coil of negligible resistance is connected in series with 90 Ω resistor across a 120 V-60 Hz line. A voltmeter reads 36 V across the resistance. Find the voltage across the coil and inductance of the coil.

Solution:

Here, R = 90 Ω, V = 120 V,

ν = 60 Hz.

$V_R = 36$ V, $V_L = ?$ $L = ?$

As $V_R = IR \therefore I = \frac{V_R}{R} = \frac{36}{90} = 0.4$ A

As $V^2 = V_R^2 + V_L^2$

$\therefore V_L = \sqrt{V^2 - V_R^2} = \sqrt{120^2 - 36^2} = 114$ v

As $V_L = I X_L$

$\therefore X_L = \frac{V_L}{I} = \frac{114}{0.4} = 286\ \Omega$

As $X_L = 2\pi vL$

$\therefore L = \frac{X_L}{2\pi v} = \frac{286}{2(3.14)\ 60} = 0.76$ H.

Example 15:

A transmitter operates at 1M Hz. The oscillating circuit has a capacitance of 200 pF. What is the inductance and capacitative reactance of the resonant circuit?

Solution:

Here $v = 1$M Hz $= 10^6$ Hz,

$C = 200$ pF $= 2 \times 10^{-10}$ F

$L = ?$ $X_C = ?$

As $v = \frac{1}{2\pi\sqrt{LC}}$

$\therefore L = \frac{1}{4\pi^2 v^2 C}$

$$= \frac{1}{4 \times \frac{22}{7} \times \frac{22}{7} \times (10^6) \times 2 \times 10^{-10}}$$

$\therefore L = 13 \times 10^{-5}$ H

As $X_C = \frac{1}{\omega C} = \frac{1}{2\pi vC}$

$$\therefore X_C = \frac{1}{2 \times \frac{22}{7} \times 10^6 \times 2 \times 10^{-10}} = 800 \text{ ohm.}$$

Example 16:

A six pole generator with fixed field excitation develops an e.m.f. of 100 V, when operating at 1500 r.p.m. At what speed must it rotate to develop 120 V?

Solution:

$$\text{New speed} = \frac{120}{100} \times 1500 \text{ r.p.m}$$

$$= 1800 \text{ r.p.m.}$$

Example 17:

A capacitor C_1 of capacitance 1 μF and another capacitor C_2 of capacitance 2μF are charged fully separately by a common battery. The two capacitors are then allowed to discharge separately through equal resistors at time $t = 0$. Which of the following is true?

(a) A current in each of the two discharging circuits is zero at $t = 0$.

(b) the current in the two discharging circuits at $t = 0$ are equal but not zero.

(c) the currents in the two discharging circuits at $t = 0$ are unequal.

(d) capacitor C_1 loses 50% of its initial charge sooner then C_2 loses 50% of its initial charge.

Solution:

During discharging of condenser, $q = q_0 e^{-t/RC}$

and $i = \dfrac{E_0 \, e^{-t/RC}}{R}$

is same in the two cases.

Also, R is same as they are discharged through equal resistors.

$\therefore$ At $t = 0$, $i = \dfrac{E_0}{R} e^0 = \dfrac{E_0}{R}$ = constant

Hence current in the two discharging circuits at t = 0 are equal but not zero.

Now, time constant, $\tau = RC$

As $C_2 > C_1$, therefore $\tau_2 > \tau_1$, or $\tau_1 < \tau_2$

i.e., C_1 would lose charge faster than C_2. Hence alternatives (b) and (d) are correct.

Example 18:

An a.c. generator consists of a coil of 50 turns and area 2.5 m^2. rotating at an angular speed of 60 rad s^{-1}, in a uniform magnetic field B = 0.30 T between two fixed pole pieces. The resistance of the circuit including that of coil is 500 Ω

(a) What is the max. current drawn from the generator?

(b) What is the flux through the coil when the current is zero? What is the flux when the current is max?

(c) Would the generator work if the coil were stationary and instead pole pieces rotated together with the same speed as above?

Solution:

(a) Here, N = 50, A = 2.5 m^2, ω = 60 rad s^{-1}

$$B = 0.30\ T,\ R = 500\ \Omega$$

Max. value of induced e.m.f.

$$e_0 = NAB\ \omega.$$

$$\therefore \quad \text{Max. current } I_0 = \frac{e_0}{R} = \frac{NAB\ \omega}{R}$$

$$= \frac{50 \times 2.5 \times 0.30 \times 60}{500} = 4.5 \text{ amp.}$$

(b) When current is zero, induced e.m.f. = 0. As induced e.m.f. is rate of change of mag. flux, therefore, mag. flux must be maximum = n AB

$$= 50 \times 2.5 \times 0.3 = 37.5 \text{ weber.}$$

On the contrary, when current is max. induced emf is max. Therefore, magnetic flux linked with the coil must be zero.

(c) Yes, the generator will work if pole pieces are rotated instead of coil. This is because only relative motion between the coil and magnet is needed.

Example 19:

An L.C. circuit contains a 20 mH inductor and a 50 mF capacitor with an initial charge of 10 μC. The resistance of the circuit is negligible. Let the instant the circuit is closed be t = 0.

(a) What is the total energy stored initially? Is it conserved during the oscillations?

(b) What is the natural frequency of the circuit?

(c) At what time is the energy stored?

(i) Completely electrical?

(ii) Completely magnetic.

(d) At what time is the total energy shared equally between the inductor and the capacitor?

(e) If a resistor is inserted in the circuit, how much energy is eventually dissipated as heat?

Solution:

Here, $L = 20$ mH

$= 20 \times 10^{-3}$ H

$C = 50$ mF $= 50 \times 10^{-6}$ F

$q_0 = 10$ mC $= 10 \times 10^{-3}$ C $= 10^{-2}$ C.

(a) $$E = \frac{q_0^2}{2C} = \frac{\left(10^{-2}\right)^2}{2 \times 50 \times 10^{-6}} = 1 \text{ joule}$$

Yes, this energy is conserved during the oscillations.

(b) $$\nu = \frac{1}{2\pi\sqrt{LC}}$$

$$= \frac{1 \times 7}{2 \times 22\sqrt{20 \times 10^{-3}\ 50 \times 10^{-6}}}$$

$$= \frac{7 \times 10^3}{44} = 159.1 \text{ hertz.}$$

(c) (i) At any instant t, charge on the capacitor is

$$q = q_0 \cos wt = q_0 \cos \frac{2\pi}{T}.t$$

At $t = 0, \frac{T}{2}, T, \frac{3T}{2},$

$q = q_0 = \max.$

∴ Energy is stored completely in the capacitor as electrical energy.

(ii) Magnetic energy around L will be maximum, when electrical energy in C is zero *i.e.*, $q = 0$.

This will be at $t = \frac{T}{4}, \frac{3T}{4}, \frac{5T}{4}$ ......

(d) Equal sharing of energy means, energy of capacitor = 1/2 maximum energy.

$$\frac{q^2}{2C} = \frac{1}{2}\frac{q_0^2}{2C}$$

i.e.,
$$q = \frac{q_0}{\sqrt{2}}$$

From $q = q_0 \cos \omega t = q_0 \cos \frac{2\pi}{T} t$

$$\frac{q_0}{\sqrt{2}} = q_0 \cos \frac{2\pi}{T} t$$

$$\cos \frac{2\pi}{T} t = \frac{1}{\sqrt{2}} = \cos (2n + 1) \frac{\pi}{4}$$

$$\therefore \frac{2\pi}{T} t = (2n + 1) \frac{\pi}{4}$$

$$t = (2n + 1) \frac{T}{8}$$

Hence, energy will be half on C and half on L at

$$t = \frac{T}{8}, \frac{3T}{8}, \frac{5T}{8} \text{..........}$$

(e) The presence of resistor in the circuit involves loss of energy. The oscillations become damped and ultimately they disappear, when total energy of 1 joule is dissipated as heat.

Example 20:

A circuit containing a 80 mH inductor and a 60 μF capacitor in series is connected to a 230 V, 50 Hz supply. The resistance in the circuit is negligible.

(a) Obtain the current amplitude and r.m.s. value.

(b) Obtain the r.m.s. values of potential drop across each element.

(c) What is the average power transferred to inductor?

(d) What is the average power transferred to capacitor?

(e) What is the total average power absorbed by the circuit?

Solution:

Here, $L = 80$ mH

$= 80 \times 10^{-3}$ H

$C = 60\ \mu F = 60 \times 10^{-6}$ F,

$R = 0$

$E_v = 230$ V, $E_0 = \sqrt{2} \times E_v$

$= \sqrt{2} \times 230$ V

$v = 50$ Hz, ω

$= 2\pi v = 100\ \pi$ rad/s

(a) $I_0 = ?\ I_v = ?$

$$I_0 = \frac{E_0}{\left(\omega L - \frac{1}{\omega C}\right)} = \frac{230\sqrt{2}}{\left(100\,\pi \times 80 \times 10^{-3} - \frac{1}{100\,\pi \times 60 \times 10^{-6}}\right)}$$

$$= \frac{230\sqrt{2}}{\left(8\pi - \frac{1000}{6\pi}\right)} = \frac{230\sqrt{2}}{-27.91}$$

$= -\ 11.63$ amp.

$$I_v = \frac{I_0}{\sqrt{2}} = \frac{-11.63}{1.414} = -\ 8.23 \text{ amp.}$$

Negative sign appears as $\omega L < \frac{1}{\omega C}$

$\therefore$ e.m.f. lags behind the current by 90°

(b) Across L, $V = I_v \times \omega L$

$= 8.23 \times 100\ \pi \times 80 \times 10^{-3}$

Across C, $V = I_v \times \frac{1}{\omega C} = 8.23 \times \frac{1}{100\pi \times 60 \times 10^{-6}}$

$V = 436.84$ volt.

As voltages across L and C are 180° out of phase, therefore, they get subtracted.

That is why applied r.m.s. voltage

$= 436.84 - 206.74 = 230.1$ volt.

(c) Average power transferred over a complete cycle by the source to inductor is always zero because of phase difference of $\pi/2$ between voltage and current through L.

(d) Average power transferred over a complete cycle by the source to the capacitor is also zero because of phase difference of $\pi/2$ between voltage and current through C.

(e) Total average power absorbed by the circuit is also, therefore zero.

Example 21:

A circuit containing resistance R_1, inductance L_1 and capacitance C_1 connected in series gives resonance at the same frequency v as a second similar combination of R_2, L_2 and C_2. If the two circuits are connected in series, show that the whole circuit resonate with the same frequency.

Solution:

Here, as the two circuits (R_1, L_1, C_1) and (R_2, L_2, C_2) resonate at the same frequency. Therefore,

$$v = \frac{1}{2\pi\sqrt{L_1 C_1}} = \frac{1}{2\pi\sqrt{L_2 C_2}}$$

$\therefore \quad L_1C_1 = L_2C_2$ or $L_1 = L_2C_2/C_1$ ...(i)

When the two circuits are connected in series,

$$L = L_1 + L_2 \text{ and } C = \frac{C_1C_2}{C_1 + C_2}$$

$$\therefore \quad LC = \frac{(L_1 + L_2)\, C_1C_2}{C_1 + C_2}$$

$$= \frac{C_1C_2}{C_1 + C_2}\left[\frac{L_2C_2}{C_2} + L_2\right] \quad \text{...Using (i)}$$

$$LC = \frac{C_1C_2L_2}{C_1 + C_2}\left[\frac{C_1}{C_2} + 1\right]$$

$$= \frac{C_1C_2L_2}{C_1 + C_2}\,\frac{(C_1 + C_2)}{C_1}$$

$LC = L_2C_2$

Hence resonance frequency of the combination

$$f' = \frac{1}{2\pi\sqrt{LC}} = \frac{1}{2\pi\sqrt{L_2C_2}} = f,$$

which was to be proved.

Example 22:

The impedance of a series RLC circuit is 8 ohm, when v = 60 Hz at resonance and 10 Ω at 80 Hz. Calculate the values of L and C.

Solution:

At resonance, $Z = R = 8\ \Omega$,

$v = 60$ Hz.

As $v = \dfrac{1}{2\pi\sqrt{LC}}$

$$\therefore\ L = \frac{1}{4\pi^2 v^2 C}$$

$$= \frac{1}{4 \times \frac{22}{7} \times \frac{22}{7} \times 60^2 C} \qquad \text{...(i)}$$

When $v = 80$ Hz, $Z = 10\ \Omega$

As $Z^2 = R^2 + (X_L - X_C)^2$

$$\therefore\ X_L - X_C = \sqrt{Z^2 - R^2}$$

$$= \sqrt{10^2 - 8^2} = 6\Omega$$

$$\omega L - \frac{1}{\omega C} = 6$$

$$2\pi\, v\, L - \frac{1}{2\pi\, v\, C} = 6$$

$$2\pi\,(80)\,L - \frac{1}{2\pi\,(80)\,C} = 6$$

Using (i), we get

$$\frac{2\pi(80)}{4 \times \frac{22}{7} \times \frac{22}{7} \times 60^2 C} - \frac{1}{2\pi(80)C} = 6$$

On simplification, we get C = 0.00027 F

Substituting this value of C in (i), we get

$$L = 0.0261 \text{ H.}$$

Example 23:

A current of 4A flows in a coil when connected to a 12 V d.c. source. If the same coil is connected to 12 volt, 50 rad/s a.c. source, a current of 2.4 A flows in the circuit. Determine the inductance of the coil. Also find the power developed in the circuit if a 2500 μF condenser is connected in series with the coil.

Solution:

Here, I = 4A, V = 12V (d.c.)

$$\therefore \quad R = \frac{V}{I} = \frac{12}{4} = 3\Omega$$

Again, E_v = 12V, Ω = 50 rad/sec,

$$I_v = 2.4\text{A}$$

$$\therefore \quad Z = \frac{E_v}{I_v} = \frac{12}{2.4} = 5\,\Omega$$

If X_L is inductive reactance of the coil, then

$$Z^2 = R^2 + X_L^2$$

$$\text{or} \quad X_L = \sqrt{Z^2 - R^2} = \sqrt{5^2 - 3^2} = 4\Omega.$$

Now, $X_L = \omega L$

$$\therefore \quad L = \frac{X_L}{\omega} = \frac{4}{50} = 0.08 \text{ H}$$

$$X_C = \frac{1}{\omega C} = \frac{1}{50 \times 2500 \times 10^{-6}}$$

$$= 10^3/125 = 8\Omega$$

Impedance of RLC circuit,

$$Z = \sqrt{R^2 + (X_C - X_L)^2}$$

$$= \sqrt{3^2 + (8-4)^2} = 5\Omega$$

$$I = E/Z = \frac{12}{5} = 2.4 \text{ A}$$

Power dissipated/cycle = I^2R

$$= (2.4)^2 \times 3 = 17.28 \text{ W.}$$

Example 24(a):

A long solenoid of diameter 0.1 m has 2 × 10⁴ turns per metre. At the centre of the solenoid is 100 turn coil of radius 0.01 m placed with its axis coinciding with the solenoid axis. The current in the solenoid is decreased at a constant rate from +2 A to –2A in 0.05 s. Find the e.m.f. induced in the coil. Also, find the total charge flowing through the coil during this time, when the resistance of the coil is $10\pi^2$ Ω

Solution:

Here, $r_1 = 0.1/2 = 0.05$ m,

$N_1 = 2 \times 10^4$

$N_2 = 100$, $r_2 = 0.01$ m

$di = -2 - 2 = -4$A, $dt = 0.5$s.

$R = 10\pi^2\ \Omega$

$e = ?\ q = ?$

Mutual inductance of the system of solenoid and coil,

$$M = \mu_0 N_1 N_2 A_2 = (\mu_0 N_1 N_2)\pi\ r_2^2$$

$$= 4\pi \times 10^{-7} \times 10^4 \times 100 \times \pi\ (0.01)^2$$

$$M = 8\pi^2 \times 10^{-5} \text{ H}$$

e.m.f. induced in the coil

$$e = \frac{M\,di}{dt} = 8\pi^2 \times 10^{-5}\,\frac{(-4)}{0.05}$$

$$= 6.31 \times 10^{-2} \text{ V}$$

Charge induced, $q = i\,dt = \dfrac{e}{R}dt$

$$= \frac{6.31 \times 10^{-2}}{10\pi^2} \times 0.05$$

$$q = 3.2 \times 10^{-5}\ C.$$

Example 24(b):

An a.c. source of angular frequency ω is led across a resistor R and a capacitor C is series. The current registered is I. If now the frequency of source is changed to $\omega/3$ (but maintaining the same voltage), the current in the circuit is found to be halved. Calculate the ratio of reactance to resistance at the original frequency ω.

Solution:

$$I = \frac{V}{\left[R^2 + \left(\frac{1}{\omega C}\right)^2\right]^{1/2}} \qquad \text{...(i)}$$

and $$\frac{I}{2} = \frac{v}{\left[R^2 + \left(\frac{3}{\omega C}\right)^2\right]^{1/2}} \qquad \text{...(ii)}$$

Divide (i) by (ii),

$$2 = \frac{\left[R^2 + (3/\omega C)^2\right]^{1/2}}{\left[R^2 + (1/\omega C)^2\right]^{1/2}}$$

Squaring, we get,

$$R^2 + \frac{9}{\omega^2 C^2} = 4\left(R^2 + \frac{1}{\omega^2 C^2}\right)$$

or $$\frac{5}{\omega^2 C^2} = 3R^2$$

$$\therefore \quad \frac{X_C}{R} = \frac{1/\omega C}{R} = \sqrt{\frac{3}{5}}\,.$$

Example 25:

An LCR series circuit with 100 Ω resistance is connected to an a.c. source of 200 volt and angular frequency 300 radians/sec. When only the capacitance is removed, the current lags behind the voltage by 60°. When only the inductance is removed, the current leads the voltage by 60°. Calculate the current and the power dissipated in LCR circuit.

Solution:

In an RLC series circuit, current lags behind the voltage by ϕ given by

$$\tan\phi = \frac{\omega L - \frac{1}{\omega C}}{R} \qquad \text{...(i)}$$

When capacitance is removed,

$$X_C = \frac{1}{\omega C} = 0$$

$$\therefore \quad \tan\phi = \frac{\omega L}{R}$$

$$\tan 60^\circ = \frac{\omega L}{R} \qquad \text{...(iii)}$$

When inductance is removed, $X_L = \omega L = 0$

$$\tan\phi = \frac{-1}{\omega CR}$$

(Neg. sign for current leading the voltage)

$$\therefore \quad \tan 60^\circ = \frac{1}{\omega CR} \qquad \text{...(iii)}$$

From (ii) and (iii), we get

$$\frac{\omega L}{R} = \frac{1}{\omega CR}$$

$$\text{or} \quad \omega L = \frac{1}{\omega C}$$

As impedance

$$Z = \sqrt{R^2 + \left(\omega L - \frac{1}{\omega C}\right)^2} = R$$

$$\therefore \quad I_v = \frac{E_v}{Z} = \frac{E_v}{R} = \frac{200}{100} = 2A$$

From (i), $\tan \theta = 0, \therefore \phi = 0^{\circ}$

$P = E_v I_v \cos \phi = (200)(2) \cos 0^{\circ} = 400$ W.

Example 26:

A uniformly wound solenoid coil of self inductance 1.8×10^{-4} H and resistance 6 Ω is broken into two identical coils. These identical coils are then connected in parallel across a 12 V battery of negligible resistance. Calculate the steady state current through the battery and time constant of the circuit.

Solution:

Here, when the solenoid is broken into two identical coils, inductance of each coil becomes half and so does the resistance.

$$\therefore \quad L_1 = L_2 = \frac{1.8 \times 10^{-4}}{2}$$

$$= 0.9 \times 10^{-4} \text{ hertz.}$$

$$R_1 = R_2 = 6/6 = 2\ \Omega$$

When these coils are connected in parallel to a battery, then

$$\text{from } \frac{1}{L} = \frac{1}{L_1} + \frac{1}{L_2} = \frac{L_2 + L_1}{L_1 L_2}$$

$$L = \frac{L_1 L_2}{L_1 + L_2} = \frac{0.9 \times 10^{-4} \times 0.9 \times 10^{-4}}{2 \times 0.9 \times 10^{-4}}$$

$$= 0.45 \times 10^{-4} \text{ H}$$

$$\text{Again, from } \frac{1}{R} = \frac{1}{R_1} + \frac{1}{R_2} = \frac{R_2 + R_1}{R_1 R_2}$$

$$\text{or} \qquad R = \frac{R_1 R_2}{R_1 + R_2} = \frac{3 \times 3}{3 + 3} = 1.5\ \Omega$$

$$\text{Time constant} = \frac{L}{R} = \frac{0.45 \times 10^{-4}}{1.5}$$

$$= 3 \times 10^{-5} \text{ s.}$$

The steady state current doesn't depend upon L.

$$\therefore \qquad I_0 = E/R = \frac{12}{1.5} = 8\text{A.}$$

Example 27(a):

Obtain the resonant frequency and Q-factor of a series LCR circuit with L = 3.0H, C = 27 μF and R = 7.4 Ω. It is desired to improve the sharpness of resonance of the circuit by reducing its full width at half max. by a factor of 2. Suggest a suitable way.

Solution:

Here L = 3.0 H

C = 27mF = 27 × 10^{-6} F

R = 7.4 Ω

Resonance frequency $\omega_r = \frac{1}{\sqrt{LC}}$ which means impedance Z of parallel LC combination becomes extremely large. Current drops to zero. The circuit behaves like an open circuit in contrast to series LCR circuit, where current becomes maximum at this frequency.

$$\omega_r = \frac{1}{\sqrt{LC}}$$

$$= \frac{1}{\sqrt{3 \times 2 \times 10^{-7}}} = \frac{10^3}{9}$$

$$= 111.1 \text{ rad s}^{-1}.$$

$$Q = \frac{\omega_r L}{R} = \frac{111.1 \times 3}{7.4} = 45.04$$

To reduce full width at half max. by a factor of 2 without changing ω_r we have to reduce the value of R to

$$\frac{R}{2} = \frac{7.4}{2} = 3.7 \text{ ohm}.$$

Example 27(b):

A horizontal copper disc of radius 10 cm rotates about a vertical axis passing through its centre with a frequency of 10 revolutions per sec. A uniform magnetic field of 100 Gauss acts perpendicular to the plane of disc. Calculate the potential difference between the centre and rim in volts.

Solution:

The magnetic flux linked with the disc ϕ = BA

($\because$ area of disc is perpendicular to the magnetic field)

According to Faraday's law, induced emf or potential difference between the centre of disc and rim is given by

$$e = -\frac{d\phi}{dt} = -\frac{d}{dt}(BA) = -B\frac{dA}{dt} = B\frac{dA}{dt} \quad \text{(numerically)}$$

Given that B = 100 gauss = 100×10^{-4} Tesla = 10^{-2} Tesla and r = 10 cm = 0.1 m

Now (dA/dt) = Area swept out by the disc/sec

$$= \frac{\text{Area swept out}}{\text{revolution}} \times \frac{\text{no. of revolutions}}{\text{sec}} = \pi r^2 f$$

$\therefore$ e = B (dA/dt) = $B.\pi r^2 f = 10^{-2} \times 3.14 \times (0.1)^2$

$\times\ 10 = 3.14 \times 10^{-3}$ volts = 3.14 mV.

Example 27(c):

Two parallel wires AL and BM placed at a distance w are connected by a resistor R and placed in a magnetic field B which is perpendicular to the plane containing the wire (see fig.). Another wire CD now connected the two wires perpendicularly and made to slide with velocity v. Calculate the work done per second needed to slide the wire CD. Neglect the resistance of all the wires.

Solution:

When the wire CD is made to slide on two rails AL and BM, with a velocity v, the magnetic flux linked with the closed circuit changes and therefore an induced emf is produced in the wire. According to Faraday's Law, induced emf is given by

$$e = -\frac{d\phi}{dt} = -\frac{d}{dt}(\vec{B}\,\vec{A}) = -\frac{d}{dt}(BA) \quad [\because \vec{B} \text{ is } \perp \text{ to area A}]$$

$$= B\frac{dA}{dt} \quad \text{(numerically)}$$

$$= B\frac{d}{dt}(wx) = wB\frac{dx}{dt} = wBv \qquad ...(1)$$

where (dx/dt) = v = velocity of wire [dx = distance moved by wire CD in time dt]

Because the circuit ABCD is closed, hence the current induced in the circuit is

$$i = \frac{e}{R} = \frac{wBv}{R} \quad \text{[R = Resistance of the wire AB]} \qquad ...(2$$

According to Lenz's Rule, this current opposes the motion of wire CD (A magnetic force acts on the wire CD in a direction opposite to the direction of its motion)

Magnitude of magnetic force acting on the wire CD carrying induced current i, is given by

$$F = Biw \sin 90^\circ = Biw = B\left(\frac{Bwv}{R}\right)w = \frac{B^2w^2v}{R}$$

As wire CD moves through a distance dx in time dt, therefore work done needed to slide the wire through a distance dx is

$$dW = F\,d\,x = \frac{B^2w^2v}{R}dx$$

∴ Work done per second needed to slide the wire Cd is

$$p = \frac{dW}{dt} = \frac{B^2w^2v}{R}\left(\frac{dx}{dt}\right) = \left(\frac{B^2w^2v}{R}\right)v = \frac{B^2w^2v^2}{R}$$

Example 28:

Two identical circular coils A and B are placed parallel to each other with their centres on the same axis. The coil B carries a current I in the clockwise direction as seen from A. What would be the direction of the induced current in A as seen from B when

(a) the current in B is increased?

(b) the coil B is moved towards A keeping the current in B constant?

Solution:

(a) When the current in coil B is increased, the magnetic flux linked with the coil Ais increased. This increase in magnetic flux will induce an emf and consequently a current in A, which by Lenz's Rule, has direction so as to oppose the increase in magnetic flux through it. Hence *the direction of the induced current in A as seen from B will be clockwise.*

(b) When the coil B is moved towards A keeping the current in B as constant, then also magnetic flux linked with coil A gets increased. Accordingly to Lenz's rule, the direction of the induced current in A will be such as to oppose the increase in flux which produces induced current. *Again the direction of the induced current in A as seen from B will be clockwise.*

Example 29:

A thin ring of a radius 10 cm carries a uniformly distributed charge. The ring rotates at a constant angular speed of 1200 revolutions per minute about its axis perpendicular to its plane. Find the charge carried by the ring if the magnetic field induction at its centre is 3.14×10^{-9} Tesla.

Solution:

The thin ring can be considered as a single loop. If it carries a charge q and if it makes N revolution/sec it is equivalent to a current of i = Nq passing through the ring.

Due to this current, the magnetic field produced at the centre of the ring will be

$$B = \frac{\mu_0 i}{2r} = \frac{\mu_0 Nq}{2r}$$

$$\therefore q = \frac{2rB}{\mu_0 N} = \frac{2 \times 0.1 \times 3.14 \times 10^{-9}}{4\pi \times 10^{-7} \times 20} = 2.5 \times 10^{-5}\ C$$

Example 30:

A square metal wire loop of side 10 cm and resistance 1Ω is moved with a constant velocity v_0 in a uniform magnetic field of induction B = 2 Weber/m^2 as shown in figure. The magnetic field lines are perpendicular to the plane of loop directed into the paper. The loop is connected to a network of resistors each of value 3Ω. The resistance of lead wires OS and PQ are negligible. What should be the speed of the loop v_0 so as to have a steady current of 1 mA in the loop? Indicate also the direction of current in the loop.

Solution:

Let us first find out the effective resistance of network of resistances. Because the given network is a balanced Wheatstone's bridge, hence points A and C will be at same potential and resistance across AC will be ineffective.

Hence Resistance of branch

$$QCS = 3\Omega + 3\Omega = 6\Omega$$

and Resistance of branch

$$QAS = 3\Omega + 3\Omega = 6\Omega$$

Now the two branches are connected in parallel between points Q and S. Hence effective resistance of the network is

$$\frac{1}{R'} = \frac{1}{6} + \frac{1}{6} \text{ or } R' \ 3\Omega$$

The resistance of the square loop in 1Ω, hence total resistance of the circuit = R = 3 + 1 = 4Ω.

The induced emf in a wire of length l moving in a uniform magnetic field with a constant velocity v_0 is given by $e = B\,v_0 l$

$$\therefore \text{ Current in the loop} = i = \frac{e}{R} = \frac{Bv_0 l}{R}$$

Hence speed of loop

$$v_0 = \frac{iR}{Bl} = \frac{10^{-3} \times 4}{2 \times 0.1} = 2 \times 10^{-2} \text{ m/s} = 2 \text{ cm/sec.}$$

Example 31(a):

A long solenoid of diameter 0.1 m has 2 × 10^4 turns per meter. At the centre of the solenoid a 100 turn coil of radius 0.01 m is placed with its axis coinciding with the solenoid axis. The current in the solenoid is decreased at a constant rate from + 2A to – 2A in 0.05 sec. Find the emf induced in the coil. Also find the total charge flowing through the coil during this time when the resistance of the coil is $10\pi^2$ ohms.

Solution:

If n is the number of turns per unit length of the solenoid and i is the current flowing through it, then magnetic induction produced at the centre of the solenoid is given by

$$B = \mu_0 \, n \, i \text{ Weber/meter}^2$$

Because magnetic induction is along the axis of the solenoid and axis of the coil (placed inside the solenoid) coincides with the solenoid axis, hence plane of coil is perpendicular to magnetic induction B. Therefore flux linked with each turn of the coil of area A, is

$$\phi = BA \text{ Weber} = \mu_0 niA \text{ Weber}$$

According to Faraday's Law of e.m. induction, induced emf produced in a coil of N turns is given

$$e = -N\frac{d\phi}{dt} = -N\frac{d}{dt}\,(m_0 \, n \, i \, A) = -N\mu_0 \, nA\frac{di}{dt}$$

Given that N = 100, n = 2 × 10^4/meter,

$$A = \pi \times 10^{-4} \text{ meter}^2$$

and $(di/dt) = [(-2-2)/0.05] = -80$ amp/sec.

$$\therefore\ e = -100 \times (4\pi \times 10^{-7}) \times (2 \times 10^4) \times (\pi \times 10^{-4}) \times (-80)$$

$$= 6.4\ \pi^2 \times 10^{-3} \text{ Volts.}$$

$$\text{Now } q = i \times t = (e/R) \times t = \frac{\left(6.4\pi^2 \times 10^{-3}\right) \times 0.05}{10\pi^2}$$

$$= 3.2 \times 10^{-5} \text{ coulomb} = 32 \text{ micro coulomb.}$$

Example 31(b):

A very small circular loop of area 5×10^{-4} m^2, resistance 2 ohms and negligible inductance is initially coplanar and concentric with a much larger fixed circular loop of radius 0.1 m. A constant current of 1 ampere is passed in a bigger loop and the smaller loop is rotated with angular velocity ω rad/sec about a diameter. Calculate (i) The flux linked with the smaller loop, (ii) induced emf, and (iii) induced current in the smaller loop, as a function of time. ($\mu_0 = 4\pi \times 10^{-7}$ V–S/A–m).

Solution:

(i) The magnetic field produced at the centre of the larger loop of radius a due to current i flowing through it, is given

$$B = \frac{\mu_0 i}{2a} = \frac{\left(4\pi \times 10^{-7}\right) \times 1}{2 \times 0.1} = 2\pi \times 10^{-6} \text{ Weber/meter}^2$$

According to Right hand palm rule no. 1, field B will be perpendicular to the plane of loop either directed upward or downward.

Hence the instantaneous magnetic flux linked with the smaller loop of area A (placed at the centre of larger loop) is

$$\phi = \vec{B}.\vec{A}. = BA \cos\theta = BA \cos\omega t \qquad [\because \omega = \theta/t]$$

$$= (2\pi \times 10^{-6}) \times (5 \times 10^{-4}) \cos\omega t$$

$$= \pi \times 10^{-9} \cos\omega t \text{ Weber}$$

(ii) Induced emf $e = -(d\phi/dt) = BA\omega \sin\omega t = \pi \times 10^{-9}\ \omega \sin\omega t$ Volts.

(iii) Induced current $i = \frac{e}{R} = \frac{\pi}{2} \times 10^{-9}\ \omega \sin\omega t$ ampere

Example 31(c):

The current in a coil of self inductance 2.0 henry is increasing according to $i = 2 \sin t^2$ ampere. Find the amount of energy spent during the period when the current changes from zero to 2 ampere.

Solution:

Given that L = 2.0 H, $i = 2 \sin t^2$

Induced emf $e = -L\,(di/dt) = -2.0\,(d/dt)$

$(2 \sin t^2) = -(4 \cos t^2).2t = +8t \cos t^2$ (Numerically)

Work done for increasing charge $d\theta$

$dW = +e(d\theta) = 8t \cos t^2\,(i\,dt) = 8\,t \cos t^2\,(2 \sin t^2)\,dt = 8t\,(2 \sin t^2 \cos t^2)\,dt = 8t \sin 2t^2\,dt$

Total work done $= \int_0 \left(8t \sin 2t^2\right) dt$

When $i = 0$, $t = 0$; When $i = 2$ amp, $\sin t^2 = 1$

or $t^2 = \pi/2$

put $2t^2 = y$ or $4\,t\,dt = dy$

$\therefore\ W = 2 \int_0^{\pi} \sin y\,dy = 2\left[-\cos y\right]_0^{\pi}$

$= -2\,[\cos \pi - \cos 0] = -2\,[-1 - 1] = 4$ Joules.

Example 32:

The given mass m of copper, drawn into a wire of radius a and formed into a circular loop of radius r, is placed perpendicular to a magnetic field B. Magnetic field B is changing in magnitude at a constant rate (dB/dt). Prove that the induced current in the loop is independent of the size of the wire or of the loop and is given by

$$i = \frac{m}{4\pi\rho d} \cdot \frac{dB}{dt}$$

where ρ is the specific resistance and d, the density of copper.

Solution:

Because magnetic field is perpendicular to the plane of loop of radius r, hence magnetic flux linked with the loop is

$$\phi = BA \cos\theta = B(\pi r^2) \cos 0° = B(\pi r^2)$$

According to Faraday's law of electromagnetic induction, induced emf produced in the loop

$$e = \frac{d\phi}{dt} = \pi r^2 \left(\frac{dB}{dt}\right)$$

$\therefore$ Induced current in the loop is given by

$$i = \frac{e}{R} = \frac{\pi r^2}{R} \cdot \left(\frac{dB}{dt}\right) \text{ [R = ohmic resistance of the loop]} \quad ...(1)$$

Now $R = \rho \frac{l}{\pi a^2}$ (a = radius of wire and l = length of wire) ...(2)

Further mass of the wire

$m = \pi a^2 \, l \, d = \pi a^2 \, (2\pi r)d$

($\because$ length of wire $= l = 2\pi r$ = circumference of loop)

$\therefore \pi a^2 = m/2\pi rd$ ...(3)

From eq. (2) & (3) we get

$$R = \rho \frac{l}{(m/2\pi\, rd)} = \frac{\rho(2\pi\, rd)(2\pi\, rd)}{m} = \frac{\rho 4\pi^2 r^2 d}{m} \quad ...(4)$$

Substituting eq. (4) in eq. (1), we have

$$i = \frac{\pi r^2 m}{\rho 4\pi^2 r^2 d}\left(\frac{dB}{dt}\right) = \frac{m}{4\pi\rho d}\left(\frac{dB}{dt}\right).$$

Example 33:

A deflection of 24 divisions is obtained in a ballistic galvanometer either by charging a capacitor of 3μF to a potential difference of 2 volts and then discharging it through the galvanometer or by connecting the ballistic galvanometer in series with a flat circular coil of 80 turns, each of radius 0.5 cm. (The total resistance of coil and galvanometer being 4000 ohms) and quickly introducing the coil into a strong magnetic field so that the plane of coil is perpendicular to the direction of magnetic field. Find the sensitivity of the galvanometer and calculate the strength of the magnetic field. The strength of the earth's magnetic field may be neglected.

Solution:

Sensitivity of a galvanometer is defined as

$$\text{Sensitivity} = \frac{\text{deflection}}{\text{charge}} = \frac{\phi}{q}$$

Given that ϕ' = 24 divisions and

$$q = CV = 3 \times 2 = 6\ \mu\ C$$

$\therefore$ Sensitivity = (24/6) = 4 divisions/μc

Now we know that $e = \frac{d\phi}{dt} = iR$

or $idt = \frac{d\phi}{R}$

$\therefore\ q = \frac{d\phi}{R} = \frac{NAB}{R}$ [dϕ = NAB = Change in magnetic flux linked with a coil consisting of N turns]

Given that $A = \pi r^2 = 25\pi \times 10^{-6}\ m^2$; N = 80 and R = 4000 ohms.

Hence, $6 \times 10^{-6} = \frac{(25\pi \times 10^{-6}) \times 80 \times B}{4000}$

$$\therefore\ B = \frac{(6 \times 10^{-6}) \times 4000}{(25\pi \times 10^{-6}) \times 80} = \frac{12}{\pi} = 3.82 \text{ Tesla.}$$

Example 34:

A long solenoid having 1000 turns per cm carries an alternating current of peak value 1 ampere. A search coil having a cross–sectional area of $1 \times 10^{-4}\ m^2$ and 20 turns is kept in the solenoid so that its plane is perpendicular to the axis of the solenoid. The search coil registers a peak voltage of 2.5×10^{-2} volts. Find the frequency of the current in the solenoid.

Solution:

If i is the current flowing through the solenoid and n is the number of turns per unit length, then magnetic field produced inside the solenoid is given by $B = \mu_0 ni$

Because magnetic field inside the solenoid is parallel to its axis and earth coil is kept perpendicular to the axis, then magnetic flux linked with a search coil of area A, consisting number of turns N, is given

$$\phi = BAN = \mu_0 ni\ AN = \mu_0 nANi_0 \sin \omega t.$$

Hence according to Faraday's Law of electromagnetic induction, induced emf is

$$e = \frac{d\phi}{dt} = \mu_0\ n\ AN\ i_0\ \omega \cos \omega t = \mu_0 n\ AN\ i_0\ 2\pi f \cos \omega t.$$

$$\therefore\ e_{peak} = \mu_0 n\ AN\ i_0\ 2\pi f$$

$$\text{or } f = \frac{e_{peak}}{\mu_0 n A N i_0 2\pi}$$

Given that $e_{peak} = 2.5 \times 10^{-2}$ volts, $\mu_0 = 4\pi \times 10^{-7}$ Wb/A–m

$n = 1000$ turns/cm $= 10^5$ turns/meter, $A = 10^{-4}$ meter^2,

$N = 20$ and $i_0 = 1$ ampere.

$$\therefore \; f = \frac{2.5 \times 10^{-2}}{\left(4\pi \times 10^{-7}\right) \times 10^5 \times 10^{-4} \times 20 \times 1 \times 2\pi} = 15.8 \text{ sec.}$$

Example 35(a):

A flat circular coil of 200 turns of radius 12.5 cm is kept on a horizontal table and is connected with a ballistic galvanometer. The combined resistance of the coil and galvanometer is 800 ohms. When the coil is suddenly rotated by an angle of 180°, the spot of light swings to a maximum reading of 30 divisions. When a capacitor of 0.1 μF charged to 6V is discharged through the same ballistic galvanometer, a maximum deflection of 20 divisions is obtained. Calculate the vertical component of earth's magnetic field.

Solution:

If B_v represents the vertical component of earth's magnetic field, then magnetic flux linked with the coil of area A and number of turns N is given by $\phi = N B_v A$

When the coil is rotated through an angle of 180°, then the magnetic flux linked with it is

$$\phi' = -N B_v A$$

$\therefore$ Change in magnetic flux

$$d\phi = \phi - \phi' = 2N B_v A.$$

Hence charge passed through the galvanometer

$$q = \frac{\text{Change in Magnetic Flux}}{\text{Resistance}} = \frac{2N B_v A}{R}$$

$$= \frac{2 \times 200 \times B_v \times \left(3.14 \times 0.125 \times 0.125\right)}{800}$$

$$= 0.024 \text{ coulomb.}$$

This charge gives a throw of 30 division in the galvanometer. If K is the ballistic constant of the galvanometer then

$q = 0.024\ B_v = 30\ K$ ...(1)

The charge on the capacitor

$q' = cv = 0.1 \times 6 \times 10^{-6} = 6 \times 10^{-7}$ Coulomb

As this charge gives a through of 20 division in the galvanometer, hence

$q' = 6 \times 10^{-7} = 20K$ ..(2)

Dividing eq. (1) by eq. (2), we get

$$\frac{0.024\ B_v}{6 \times 10^{-7}} = \frac{30}{20}$$

$$\text{or } B_v = \frac{3}{2} \times \frac{\left(6 \times 10^{-7}\right)}{0.024} = 0.37 \times 10^{-4} \text{ Weber/meter}^2.$$

Example 35(b):

A heart pacing device consists of a coil of 50 turns and radius 1 mm just inside the body with a coil of 1000 turns and radius 2 cm placed coaxially just outside the body. Calculate the induced emf produced in the internal coil if a current of 1 amp in the external coil collapses in 10 milliseconds.

Solution:

Induced emf produced in the internal coil due to change in current flowing in the external coil, is given by

$$e = -\frac{d\phi}{dt} = -\frac{d}{dt}(N_2A_2B_1) = N_2A_2\frac{dB_1}{dt}$$

Where B_1 = Magnetic field produced in the external coil

$$= \frac{\mu_0 N_1 i}{2r_1}$$

$$\therefore \frac{dB_1}{dt} = \frac{\mu_0 N_1}{2r_1}\left(\frac{di}{dt}\right) = \frac{\left(4\pi \times 10^{-7}\right) \times 10^3}{2 \times \left(2 \times 10^{-2}\right)}$$

$$\left(\frac{-1}{10 \times 10^{-3}}\right) = -\pi \text{ Tesla/sec}$$

Hence $e = -(50) \times \pi\ (10^{-3})^2 \times (-\pi)$

$= 50\pi^2 \times 10^{-6}$ volt.

Example 35(c):

A circular coil P of 100 turns and radius 2 cm is placed coaxially at the centre of another circular coil Q of 1000 turns and radius 20 cm, calculate

(a) the mutual inductance of the coils

(b) the induced emf in the coil P when the current in coil Q decreases from 5 amp to 3 amp in 4 sec.

(c) the rate of change of flux through coil P at this instant and

(d) the charge passed through coil P if its resistance is 8 ohms.

Solution:

(a) The magnetic field produced at the centre of coil Q is given by

$B_1 = (\mu_0 N_1 i_1)2r_1$ ...(1)

According to Right hand palm rule no. 1, this field will be along the axis of coil Q. Because coil P is placed coaxially inside the coil Q hence magnetic field B_1 will be perpendicular to plane of coil P. Hence magnetic flux linked with coil P is given by

$$\phi_2 = N_2 B_1 A_1 = N_2 \left(\frac{\mu_0 N_1 i_1}{2r_1}\right) A_2 \quad ...(2)$$

According to definition of mutual inductance $\phi_2 = M\, i_1$

$$\therefore \text{ Mutual inductance } M = \frac{\mu_0 N_2 N_1 A_2}{2r_1}$$

or $$M = \frac{\left(4\pi \times 10^{-7}\right) \times 1000 \times 100 \times \pi\left(2 \times 10^{-2}\right)^2}{2 \times \left(20 \times 10^{-2}\right)}$$

$$= 39.44 \times 10^{-3} \text{ H} = 39.44 \text{ mH.}$$

(b) Induced emf e = M (di/dt) (Numerically)

$$= 39.44 \times 10^{-3} \times \left(\frac{5-3}{4}\right) = 19.72 \times 10^{-3} \text{ V} = 19.72 \text{ mV.}$$

(c) The rate of charge of flux through coil P

$$\frac{d\phi}{dt} = e = 19.72 \times 10^{-3} \text{ Weber/sec} = 19.72 \text{ milli Weber/sec.}$$

(d) The change passing through the coil P is

$$q = \int_0 i\,dt = \int_0 \frac{e}{R}\,dt = -\int_0 \frac{M\left(di/dt\right)dt}{R}$$

$$= -\frac{M}{R}\int_{i_1}^{i_2} di = \frac{M}{R}(i_1 - i_2) = \frac{39.44 \times 10^{-3}}{8}$$

$$(5 - 3) = 9.86 \times 10^{-5} \text{ coulomb} = 98.6 \ \mu C.$$

Example 36:

The capacitor of an oscillatory circuit of negligible resistance is enclosed in a container. When the container in evacuated, the frequency of the circuit is 150 kcs^{-1}, and when the container is filled with a gas, the frequency changes by 100 c/s. Find the dielectric constant of the gas.

Solution:

Here $v = 150 \text{ kc s}^{-1}$

$= 150 \times 10^3 \text{ cs}^{-1}$

$v' = (150 \times 10^3 - 100) \text{ cs}^{-1} = 149900 \text{ cs}^{-1}$

Now $v = \dfrac{1}{2\pi\sqrt{LC}}$ and $v' = \dfrac{1}{2\pi\sqrt{LKC}}$

Dividing, we get $\dfrac{v}{v'} = \sqrt{K}$

or $K = \left(\dfrac{v}{v'}\right)^2 = \left(\dfrac{150 \times 10^3}{149900}\right)^2$

$K = 1.0012.$

Example 37:

Two circuits A and B connected by identical d.c. sources (e.m.f. = 10V) differ greatly in their self-inductance. Circuit A has large self-inductance = 10H, while the self-inductance of B is much smaller equal to 0.05 H. The total external resistance in each circuit (which includes the resistance of the inductor itself) is 40 Ω

(a) *Are the steady current values in each circuit equal? If so, what is the value?*

(b) *Compare the times required by the currents in the two circuits to reach* $\left(1 - \dfrac{1}{e}\right)$ *of their steady value?*

(c) *Which circuit requires greater energy consumption of the source to build up its current to the steady value?*

(d) *After the steady state is reached, do the circuits dissipate the same power in the form of heat?*

Solution:

(a) Here, $E = 10V$, $L_A = 10H$,

$L_B = 0.05$ H, $R = 40\ \Omega$

Yes, the steady current values in the two cases are same.

$$I_0 = \frac{E}{R} = \frac{10}{40} = 0.25 \text{ amp.}$$

(b) Relation governing growth of current in RL circuit is

$$I = I_0\ [1 - e^{(-R/L)\,t}]$$

In $t = \frac{L}{R}$, $I = I_0\ [1 - e^{(-R/L \times L/R)}]$

$$= I_0 \left(1 - \frac{1}{e}\right)$$

i.e., current reaches $\left(1 - \frac{1}{e}\right)$ of the steady value in

$$t = \frac{L}{R}$$

Now, $t_A = \frac{L_A}{R}$, $t_B = \frac{L_B}{R}$

$$\therefore \quad \frac{t_A}{t_B} = \frac{L_A}{L_B} = \frac{10}{0.05} = 200$$

(c) Energy required to build-up current I_0

$$E = \frac{1}{2} L\, I_0^2$$

$$\therefore \quad E_A = \frac{1}{2} \times 10\ (0.25)^2 = 0.31 \text{ joule}$$

$$E_B = \frac{1}{2} \times 0.05\ (0.25)^2$$

$$= 1.55 \times 10^{-3} \text{ joule.}$$

(d) As power dissipated $= \frac{E^2}{R}$ (No dependance on L)

$\therefore$ It is same in two cases.

Example 38:

Two concentric coplanar circular loops made of wire, with resistance per unit length $10^{-4}\ \Omega m^{-1}$, have diameters 0.2m and 2m. A time varying potential difference (4 + 2.5t) volts is applied to the larger loop. Calculate the current in the smaller looper.

Solution:

Given that the Resistance per unit length of the wires of each circular loop = $\rho = 10^{-4}$ ohms/m R = Radius of outer loop = 1m, r = Radius of inner loop = 0.1 m

Magnetic field at the common centre O, due to current in larger loop, is given by

$$B_0 = \frac{\mu_0 i}{2R} = \frac{\mu_0}{2R} \times \frac{\text{Potential difference}}{\text{Resistance}}$$

$$= \frac{\mu_0}{2R} \times \frac{(4+2.5t)}{(2\pi R)\rho} = \frac{\mu_0}{2R}\left(\frac{4+2.5t}{2\pi R\rho}\right)$$

Because the inner loop is very small (r << R) hence field at every point inside it can be assumed as constant (= B_0).

Thus flux linked with smaller loop,

$$\phi = B_0 \pi r^2 = \frac{\mu_0}{2R}\left(\frac{4+2.5t}{2\pi R\rho}\right)\pi r^2$$

Hence, induced e.m.f. $|e| = +\frac{d\phi}{dt} = \frac{\mu_0 \rho^2}{4R^2\rho}(2.5)$

$\therefore$ current in smaller loop $= \frac{|e|}{\text{Resistance}}$

i.e., $$i' = \frac{2.5\mu_0 r^2}{4R^2\rho} \times \frac{1}{2\pi r\rho}$$

$$= \frac{2.5 \times 4\pi \times 10^{-7} \times 0.1}{4 \times 1 \times 10^{-8} \times 2\pi} = 1.25 \text{ amp.}$$

Example 39:

A closed coil consists of 500 turns on a rectangular frame of area 4.0 cm² and has a resistance of 50 ohms. The coil is kept with its plane perpendicular to a uniform magnetic field of 0.2 Weber/meter². Calculate the amount of charge flowing through the coil if it is turned over (rotated through 180°). Will the answer depend on the speed with which coil is rotated?

Solution:

Because magnetic field B is perpendicular to the plane of coil, hence magnetic flux linked with each turn of the coil of area A, is given by $\phi_1 = BA$

When the coil is rotated through 180°, magnetic flux linked with it is given by

$$\phi_2 = BA \cos 180^\circ = -BA$$

Hence change in magnetic flux

$$\phi = \phi_2 - \phi_1 = -BA - (+BA) = -2BA$$

According to Faraday's 2nd law, induced emf produced in a coil consisting of N turns, is

$$e = -N\frac{d\phi}{dt} = \frac{2\ NBA}{dt}$$

where dt represents the time interval during which change in flux $d\phi$ occurs.

If R is the resistance of the coil, then current induced in the coil is

$$i = \frac{e}{R} = \frac{2\ NBA}{R\,dt}$$

Hence charge passed through the coil in time dt is

$$q = idt = \frac{2\ NBA}{A}$$

$$= \frac{2 \times (500) \times 0.2 \times \left(4 \times 10^{-4}\right)}{50} = 16 \times 10^{-4} \text{ coulombs.}$$

Induced charge will remain same whether the coil is rotated slowly or rapidly because it depend upon the total change in magnetic flux but not on the rate of change of magnetic flux.

Example 40:

A 0.1 meter long conductor carrying a current of 50 amp lies perpendicular to magnetic field of 1.25 mT. Find out (a) the force acting on the conductor, (b) mechanical power required to move this conductor against this force with a speed of 1 m/s and (c) the emf induced in the conductor.

Solution:

(a) Mechanical Force exerted on the conductor due to magnetic field

$$F = B\ I\ L \sin 90^\circ = (1.25 \times 10^{-3}) \times (50) \times (0.1)$$
$$= 6.25 \times 10^{-3}\ N.$$

(b) Mechanical power required to move this conductor against this force with a speed of 1 m/s.

$$P = Fv = (6.25 \times 10^{-3}\ N)\ (1\ m/s) = 6.25\ mW.$$

(c) E.M.F. induced in the conductor

$$e = BvL = (1.25 \times 10^{-3})\ (1.0) \times (0.1) = 1.25 \times 10^{-4}\ volt.$$

Example 41:

A conducting rod MN of length L = 0.5 m and resistance r = 0.5Ω slides without friction with a constant velocity v = 2m/s over two conducting rails PQ and ST lying in the plane of paper. The whole system is present in a magnetic field of induction B = 0.6 Tesla which is perpendicular to the plane of paper and directed downwards. As resistance R = 2.5 Ω connected points Q and S. Calculate :

(a) the current flowing in the circuit

(b) the force in the direction of motion to be applied to the conductor for the later to move with velocity v

(c) the thermal power dissipated in the circuit. Neglect the resistance of rails PQ and ST.

Solution:

Because the conducting rod MN moves with a velocity v perpendicular to uniform magnetic field of induction B, hence induced emf produced between its two ends

$$E = BvL$$

Total resistance of the circuit = (R + r)

(a) Hence current flowing in the circuit

$$i = \frac{BvL}{R+r} = \frac{(0.6)\,(2.0)\,(0.5)}{(2.5+0.5)} = 0.2\ \text{amp.}$$

(b) Force required to maintain the motion of the motion of conductor = Force exerted on the conducting rod carrying current i due to magnetic field

$$F = BiL = \frac{B^2L^2v}{R+r} = \frac{(0.6)^2 \times (0.5)^2 \times 2}{(2.5+0.5)} = \frac{(0.6)^2}{3 \times 2} = 0.06\ N.$$

(c) Thermal power dissipated in the circuit

$= Fv = 0.06 \times 2 = 0.12W.$

EXERCISES

1. In an oscillatory circuit L = 0.2 henry and C = 0.0012 microfarad. What is the maximum value of resistance for the circuit to be oscillatory?
2. Given a coil of self-inductance 5 mH and resistance 0-5 ohm, calculate the value of the capacity of the condenser required to generate the frequency of 1 ke/sec.
3. Find whether the discharge of a condenser through the following inductive circuit is oscillatory. C = 0.1 microfarad, L = 10 mH, R = 200 ohms. If so, calculate frequency.
4. If a condenser of 10 microfarad be charged to a potential difference of 50 volts and then discharged through a coil of negligible resistance and inductance 10 mH. Find the frequency and maximum amplitude of the resultant current oscillations.
5. A coil of resistance 11 ohms and inductance 0.1 henry is connected to a 110 volts D.C. mains. Find (a) the current finally established in the coil, (b) the voltage used in overcoming the resistance when the rising current is 3 amp. (c) the rate at which the current is rising at that instant.
6. A coil having resistance of 150 ohms and an inductance of 10 henries is connected to a 90 volts supply. Determine the value of current after 2 sec.
7. An inductor of self-inductance 500 millihenry and resistance of 5 ohms is connected to a battery of negligible internal resistance. Calculate the time in which the current will attain half of its final steady value.
8. A coil of resistance 20 ohms and inductance 0-5 henry is switched in direct current 200 volts supply. Calculate the rate of increase of current
   (1) at the instant of closing the switch and
   (2) at t = L/R sec. after the switch is closed, also

   Find the steady value of the current in the circuit.
9. Obtain an expression for energy density in a magnetic field.

10. A rectangular coil of 100 turns and area 400 sq. cm. is rotated about a horizontal axis at a constant rate of 60 revolutions per sec. in a horizontal magnetic field of 2000 oersted perpendicular to the axis. Calculate (1) the maximum value of the induced emf in the coil, (2) the induced emf when the plane of the coil is 30° to the horizontal and (3) the induced emf when the place of the coil is vertical.

11. When a copper disc of radius 10 cm makes 20 rotations per sec with its plane perpendicular to a uniform magnetic field an induced emf of 3-14 millivolts is generated between the centre and the edge of the disc. Calculate B.

12. A coil has an inductance of 0.03 henry. Calculate the emf induced when a current in the coil changes at a rate of 200 amperes per second.

13. An air core solenoid of 80 cm length has 600 turns and its circular cross-section has a diameter of 2-00 cm. Calculate

    (a) The self-inductance of the solenoid

    (b) The self-linked flux when the current in the solenoid is 2 amp.

    (c) The rate of change of current in the solenoid that will produce a self-induced emf of 0-3 volt.

14. A solenoid having a core of 10-3 sq. metre its cross-section, half air and half iron is 1 metre long. If the number of turns on it is 5000, what will he its coefficient of self-inductance ?

15. An emf of 25 millivolts is induced in a coil when the current in the neighbouring circuit changes by 10 amperes in 0.1 sec. What is mutual inductance of the coil?

16. A long solenoid of length 1 metre, cross-section 10 cm^2 having 100 turns, has wound over its centre a small coil of 20 turns. Find the mutual inductance of this system.

17. Two coils, a primary of 600 turns and a secondary of 30 turns are wound on an iron ring of mean radius 10 cm and cross-section 4 cm. diameter. Find their mutual inductance for iron $\mu = 800$.

18. What energy is stored in a magnetic field of solenoid of inductance 5 milli henry when maximum current of 3 amp flows through it?

19. An inductor of 3 henries and resistance 0 ohms is connected to the terminals of a battery of emf, 12 volts and of negligible internal resistance. Calculate.
    (i) the initial rate of increase of current in the circuit.
    (ii) the rate of increase of current at the instant when the current in the circuit is 1 amp.
    (iii) the instantaneous value of current at 0-2 sec. after the circuit is closed,
    (iv) the final steady-state current
    (v) the power input to the inductor at the instant when the current is 0.5 amp,
    (vi) the rate of development of heat at this instant.
    (vii) the rate at which the energy of magnetic field is increasing at this instant,
    (viii) The energy stored in the magnetic circuit when the current has attained its steady value.
20. A capacitor of capacity 1 microfarad is dischargd through a high resistance. The time taken for half the charge in the capacitor to leak is found to be 10 sec. Compute the value of resistance.
21. Describe with full theory the method of measuring high resistance by the method of leakage.
22. An electric circuit consists of a resistance, inductance and a condenser. Discuss the growth of current in the circuit when an emf is applied.
23. Give the theory of oscillatory discharge of a condenser through a circuit containing an inductance and a resistance.
24. Explain what is meant by coefficient of mutual inductance? Describe a method of finding it experimentally.

2

Magnetization

INTRODUCTION

If a specimen is placed in a magnet field, the field in the vicinity of the sample is changed. How much it is changed depends on the material and on the shape of the sample, but for most specimens it is not altered very much-typically only by about one part in 10^5. For certain materials, however, the field at some points nearby can be increased by a factor of one hundred or so over its value in the absence of the specimen. Such materials are called ferromagnetic. Non-ferromagnetic materials which produce only a small change in field, can be either diamagnetic or paramagnetic. If the field is decreased very slightly then the material is diamagnetic. If the field is increased slightly then the material is paramagnetic. The most easily observed characteristics of the different materials are their behaviours in strong non-uniform fields. A ferromagnetic material is strongly pulled into the field. A diamagnetic material is repelled towards the field tree regions, but so weakly that sensitive apparatus is required to measure the repulsive force A paramagnetic material is pulled into the field, although the attraction is usually weak. This attraction or repulsion can be explained in terms of the interaction of magnetic dipole moments with the field in which the materials are placed. Certain atoms or molecules may have permanent magnetic dipole-moments. If they do have permanent moments, the material is paramagnetic, other wise it is diamagnetic. If paramagnetic molecules (or materials) are placed in a magnetic field there is a net alignment of the dipoles in the field direction and the substance becomes magnetised.

In a non-uniform magnetic field a para-magnetic material will thus experience a force in the direction of increasing field. All atoms and molecules acquire an induced magnetic dipole moment when placed in a magnetic field. Induced magnetic dipole moments, unlike induced electric dipole-moments, are in a direction opposite to the field in the specimen. Thus, if the molecules of a substance have no permanent moments *i.e.,* the substance is diamagnetic, a sample inserted into a non-uniform magnetic field will experience a force in the direction of decreasing field strength. Clearly, the process of alignment of the dipoles as well as the phenomena by which atones or molecules acquire an induced magnetic deployments when a substance is placed in a magnetic field is known as magnetisation. Thus, any process by which a substance acquires (or develops) a net magnetic moment is magnetisation. If we want to know the origin of diamagnetism and paramagnetism we must know the origin of magnetic dipole-moment of an atom or molecule. There are two chief contributions to the magnetic dipole-moment of an atom or molecule.

ATOMIC CURRENTS OR AMPERIAN CURRENTS

A careful study of the various magnetic phenomena associated with matter indicates that magnetism is an atomic or molecular property and that there is no intrinsic magnetic substance. In the light of this, the hypothesis is adopted that all magnetic properties of matter can be explained in terms of the motions of the electrons and positive charges associated with its constituent atoms and molecules.

All matter is made of atoms each of which consists of electrons those move in the orbits around the nucleus and rotate or spin about their own axe. These circulating electrons constitute a current without a charge transport, we call this as "Atomic currents" and is different from true-current which corresponds to charge transport. Such circulating currents were first postulated by Ampere, and are often called Amperian currents.

ABSENCE OF MONO-POLE (OR GAUSS'S LAW OF MAGNETISM)

The relation div $\vec{B} = 0$ (everywhere) is mathematically representing the fact that lines of $\vec{B}$ are closed on themselves, whenever the divergence of vector point function is zero, it means no sources or sinks exist in the region over which the divergence is zero.

In the same way Gauss's law is a formal way of stating a fundamental fact of magnetism, *i.e.,* the isolated magnetic pole does not exist, the law (we have already discussed) states that the magnetic flux through any closed surface is zero, *i.e.,*

$$\phi_B = \oint \vec{B}.d\vec{S} = 0 \qquad ...(1)$$

where integral is to be taken for entire surface. In contrast for electric field, we have

$$\varepsilon_0 \oint \vec{B}.d\vec{S} = q \text{ or div } \vec{E} = \frac{\rho}{\varepsilon_0} \qquad ...(2)$$

By the comparison of the two equations A and B it is evident that in magnetism there is no counterpart to the free charge q.

THEORIES OF MAGNETISM

The fields of permanent magnets (outside the magnets) may be described either in terms of equivalent currents or equivalent poles*. Since the entire description of magnetism has been based on the assumption that are produced by moving charges, we are led to believe that the interpretation of the field of a permanent magnet in terms of electric charges giving rise to atomic currents is a more fundamental one than the concept of magnetic charges. The electrons have at least two motions in an atom giving rise to two kinds of currents listed below:

1. The currents due to orbital motion of the electrons in atoms.
2. The currents due to spinning of electrons about their own axes.

Such circulating currents were first postulated by Ampere, and are often called amperian currents.

The question now arises as the which is more important, the orbital motion of the electrons or the electronic spin. The electronic spin takes precedence in an explanation of the behaviour of magnetic materials.

In many atoms the electrons are completely paired, *i.e.,* for each electron spinning in one direction there is an electron spinning in the opposite direction. And the same situation exists in regard to the orbital motion of the electrons. Thus, the net current circulating about any given axis is zero and there are no amperian currents in such substances. Such materials exhibit very weak magnetic effect, they are called "non-magnetic". In other words, the electron spins and orbital motions are exactly balanced out, so that any particular atom has no average amperian current. Such substances when placed in magnetic field little extra amperian currents are generated inside the atom by induction, due to the changes in the angular velocity of the electrons in their orbits. According to Lenz's law, these currents are in such a direction as to oppose the applied field. This effect is known as "diamagnetism".

In a magnetic substance, however, the cancellation is incomplete and the amperian currents are not zero. Obviously, those substances whose atoms have an odd number of electrons must have at least one uncompensated spinning electron. These electrons have the same effect as tiny electric currents known as amperian currents.

In other words, a magnetic material contains tiny amperian currents distributed throughout its entire volume. *Magnetisation consists in the alignment of these amperian currents.* If the magnetic field due to this amperian current has the same direction as the applied magnetic field, obviously the field is increased by the presence of the amperian current. This magnetic effect is known as "paramagnetism".

However, the paramagnetism arising from the spins of the electrons responsible for conduction in a metal constitutes an exception, we will not be discussing the phenomena here. Paramagnetism is generally fairly weak because lining up forces are relatively small compared with the forces from the thermal motions which try to change the order.

In the case of iron there are as many as four uncompensated spinning electrons within each atom of molecule and consequently iron tends to be "strongly magnetic". Substances that behave like iron are called "ferromagnetic".

Langevin Theory of Diamagnetism

Let us now examine how the motion of the electron in its orbit is affected by an applied field $\vec{B}$ perpendicular to the plane of the orbit. If A is the area of the orbit, the magnetic flux is BA. According to faraday's law of electro-magnetic induction induced emf ξ is given by,

$$\xi = \frac{d\phi}{dt} = A\frac{dB}{dt} \quad ...(1)$$

The emf change the motion of the electrons in their orbits. Corresponding to this induced emf, there is an electric field which acts tangentially to the direction of the motion of the electrons. The line integral of this field E around the orbit is equal to the induced emf.

$$\xi = \oint \vec{E}.d\vec{l} = E\oint dl = 2\pi rE$$

or $$E = \frac{\xi}{2\pi r}$$

where r is the radius of the orbit. The electrons are accelerated by this field and experience a force eE. Hence by Newton's law,

$$m_e \frac{dv}{dt} = eE = \frac{e\xi}{2\pi r}$$

and x is given by equation (1), hence

$$m_e \frac{dv}{dt} = \frac{e}{2\pi r}\frac{d\phi}{dt} \qquad ...(2)$$

where me is mass of the electron and v is velocity. Consider, during the time flux changes from 0 to f, the velocity of electron changes from v_o to $v_o + \Delta v$ integrating the equation (2), we get,

$$m_e \int_{v=v_0}^{v=v_0+g\Delta v} \frac{dv}{dt} dt = \frac{e}{2\pi r}\int_0^{\phi} \frac{d\phi}{dt} dt$$

or $$m_e \left|v\right|_{v_0}^{v_0+\Delta v} = \frac{e}{2\pi r}\left|\phi\right|_0^{\phi}$$

or $$\Delta v = \frac{e}{2\pi m_e r}\phi \qquad ...(3)$$

This change Δv gives rise to change in angular velocity of the electron Thus, the change in angular velocity of the electron,

$$\Delta\omega L = \frac{\Delta v}{r} = \frac{e}{2\pi m_e r^2}\phi$$

But $\phi = BA$ and $\pi r^2 = A$, where A is the area of the orbit, we get,

$$\Delta\omega L = \frac{eBA}{2m_e A} = \frac{e}{2m_e}B \qquad ...(4)$$

This change in angular velocity is called "*Larmar angular velocity*" or in terms of frequency "*Larmar frequency*". However, the change in angular velocity acts in sense opposite to that of electron.

Larmar has shown that the effect of the applied field is a "precessional motion" of the orbit without change in its form. The centripetal force necessary to hold the electron in its orbit is given by,

$$F = \frac{m_e v_o^2}{r} = m_e r \omega_o^2$$

where ω_o is the angular frequency of the electron in the orbit without magnetic field. When a magnetic field is applied the angular frequency changes by ΔωL. *The condition of stable motion in the same orbit i.e., if electron is still in the orbit of the same radius, its centripetal force must increase correspondingly i.e., the centripetal force.*

$$F' = m_e r\,(\omega_o + \Delta\omega_L)^2$$

$$= m_e r\omega_o^2 + 2m_e r\omega_o\,\Delta\omega_L + m_e r\,\Delta\omega_L^2$$

Assuming that $\Delta\omega_L < \omega_o$ and this corresponds to small value of B thus, term with $\Delta\omega_L^2$ can be neglected. Hence,

$$F' = m_e r\omega_o^2 + 2m_e r\omega_o\,\Delta\omega_L \quad ...(5)$$

when a magnetic field is applied an additional force acts on the electron.

The magnitude of this force is given by,

$$FB = evB$$

or $$FB = er\,(\omega_o + \Delta\omega_L)\,B \quad ...(6)$$

Since $v = \omega_r$ and $\omega = \omega_o + \Delta\omega_L$, FB is perpendicular to both $\vec{v}$ and $\vec{B}$. If electron is moving in anti-clockwise direction and $\vec{B}$ directed out from the orbit, the magnetic force FB will be directed towards the centre of the orbit (*i.e.*, $\vec{B}$ along z-axis $\vec{v}$ along negative y-axis hence $F_B = e(\vec{v} \times \vec{B})$ is along + x-axis). If electron is moving in clockwise direction then this force is outward as shown in Fig. 2.18.

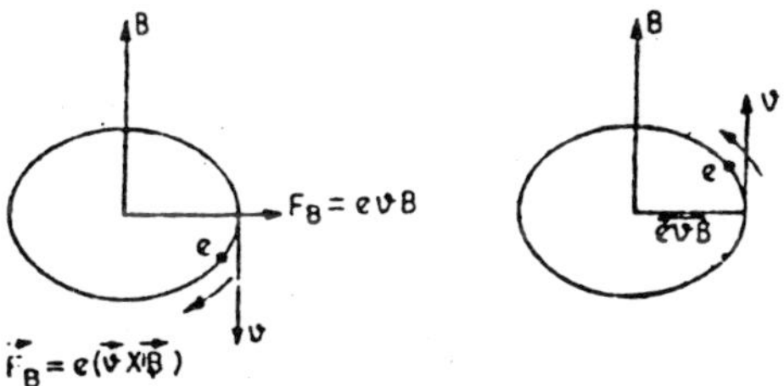

Fig. 2.1

Putting the value of B from equation (5) in eqn. (6),

$$F_B = er\,(\omega_o + \Delta\omega_L)\,\frac{2m_e}{e}\,\Delta\omega_L$$

$$= 2m_e r\omega_o\,\Delta\omega_L + 2m_e r\,\Delta\omega_L^2$$

The term $\Delta\omega_L^2$ is neglected because of the same reasons. Thus,

$$F_B = 2m_e r\omega_o \,\Delta\omega_L \qquad ...(7)$$

By equation (5), the increases in centripetal force,

$$F' - m_e r\omega_o^2 = 2m_e r\omega_o \,\Delta\omega_L \qquad ...(8)$$

The comparison of equations (7) and (8) shows that the increase in centripetal force is completely balanced by the additional inward magnetic force. Hence the centripetal *force does not change by the application of B. Therefore, the radius of the orbit will remain the same.*

A charge e revolving in a circle at the rate of n revolutions per second is "*equivalent*" to a current *ne*. If A is the area of the circle the "*magnetic moment*" of the electron orbit is given by,

$$m = neA, \text{ (since } m = IA)$$

The change in the magnetic moment dm due to change $\Delta\omega_L$ in angular velocity is given by,

$$dm = eA\Delta n. \text{ since } n = \frac{\omega}{2\pi}, \; \Delta n = \frac{\Delta\omega_L}{2\pi}$$

$$= eA\frac{\Delta\omega_L}{2\pi}$$

Substituting $\Delta\omega_L$ from eqn. (5), we get,

$$dm = eA\frac{eB}{4\pi m_e}$$

$$= \frac{e\pi r^2 eB}{4\pi m_e} \text{ (since } A = \pi r^2)$$

$$= \frac{e^2 r^2}{4m_e} B \qquad ...(9)$$

But diamagnetism is an induced effect. This magnetic moment is induced and is always opposite to $\vec{B}$, hence in vector form we can write,

$$d\vec{m} = -\frac{e^2 r^2}{4m_e}\vec{B} \qquad ...(10)$$

We have derived this equation (10) for one electron. It can be extended to an atom having more than one electron. However, in deriving this equation, we have assumed that the orbit is perpendicular to the field. Since the orbit can have any orientation with respect to the field, the radius of the orbit in this equation should be replaced by its projection r_1 on the plane perpendicular to the field, *i.e.*,

$$d\vec{m} = -\frac{e^2 r_1^2}{4m_e}\vec{B} \qquad ...(11)$$

As there are number of randomly oriented orbits within the atom, thus the total magnetic moment induced,

$$\vec{M}_A = -\frac{e^2\vec{B}}{4m_e}\Sigma r_1^2 \qquad ...(12)$$

where summation is taken for all orbits. Suppose (x, y, z) are the coordinates of any point on an orbit of radius r, then,

$$r^2 = x^2 + y^2 + z^2$$

If we choose z-axis in the plane perpendicular to the applied magnetic field, then

$$r_1^2 = x^2 + y^2$$

Let the mean value of the radii of all orbits be r,

$$r^2 = x^2 + y^2 + z^2$$

Fro spherically symmetrical atom,

$$x^2 = y^2 = z^2 = \frac{1}{2}r^2$$

hence, $$r_1^2 = x^2 + y^2 = \frac{2}{3}r^2$$

Substituting the value of r_1^2 in eqn. (12), we get,

$$\vec{M}_A = -\frac{e^2\vec{B}}{6m_e}\Sigma r^2 \qquad ...(13)$$

For a substance containing n molecules per unit volume, the total magnetic moment induced per unit volume of the substance,

$$\vec{M}_v = -\frac{e^2 n\vec{B}}{6m_e}\Sigma r^2 \qquad ...(14)$$

By equation (14), the susceptibility,

$$\chi_m = \frac{\text{Magnetic moment}}{H}$$

The diamagnetic susceptibility per atom is called Atomic susceptibility denoted by χ_{mA}. Then,

$$\chi_{mA} = \frac{\text{Induced magnetic moment per atom } M_A}{H}$$

Substituting MA from equation (14), we get,

$$\chi_{mA} = -\frac{e^2 B}{6m_e H}\Sigma r^2$$

For diamagnetic substance, we have linear relation,

$$B = \mu H$$

Hence, $$\chi_{mA} = -\frac{\mu e^2}{6m_e}\Sigma r^2 \quad ...(15)$$

Similarly, the volume susceptibility,

$$\chi_{mv} = -\frac{n\mu e^2}{6m_e}\Sigma r^2 \quad ...(16)$$

The scrutiny of the above results leads to the following conclusions:

1. The expression for susceptibility is independent of temperature.
2. The diamagnetism arises from the effect of magnetic field on the motion of electrons in the orbit and every substance having moving electrons in the orbit. Hence, every substance possesses diamagnetism, but diamagnetism is a feeble effect and overridden by the paramagnetism and ferromagnetism producing a flux in the same direction as that externally applied field.
3. The negative sign indicates that diamagnetism is an induced opposing effect, that disappears as soon as the field is removed.

LANGEVIN'S THEORY OF PARAMAGNETISM

As we have pointed earlier that the molecules or atoms with permanent magnetic dipole moment are termed paramagnetic. For such molecules there is atomic (or amperian) current which gives rise to the magnetic dipoles having magnetic dipole moment. They have amperian current, because the orbit of the electrons are so oriented that the amperian currents due to their orbital motion as well as spin do not cancel out completely. This happens with all the atoms and molecules having odd number of electrons for them the total spin is not zero. In most bulk materials, there is a net magnetic moment only if there are atoms present whose inner electron shell is no filled.

Langevin theory of paramagnetism is just like the Langevin theory of dielectrics (polar-molecules). Suppose each atom or molecule possesses a permanent magnetic moment m. When an external magnetic field $\vec{B}$ is applied, the *dipoles align along the direction of applied field.* Let the axis of dipole be making an angle θ with the direction of applied field B. The magnetic potential energy is given by,

$$U = -\vec{m}.\vec{B}$$

$$= -mB\cos\theta$$

However, the alignment of dipoles will be counteracted by the collisions of thermal motion between the molecules which knock the axes out of line. As a result, there will be an equilibrium distribution of the axes with reference to the direction of the field, which can be determined with the help of the kinetic theory of gases as follows.

According to the kinetic theory principle, the number of molecules distributed with the axis of the dipole pointing in all directions within a solid angle Ω at an angle θ with a given line of reference is proportional to dΩ. And on the basis of Boltzmann theorem, the number of molecules whose potential energy is v is proportional to $e^{-U/KT}$. Considering unit volume of the gas, if dN out of the total number N of the dipoles in it have their magnetic axes inclined at the angle π with the field direction, then, applying the principles of kinetic theory states above, we get,

$$dN \propto e^{-U/KT}\, d\Omega$$

$$= C\, e^{-U/KT}\, d\Omega$$

where C is constant depending on the number of molecules considered, U is potential energy, K is Boltzmann constant and T is absolute temperature. We have calculated that

$$d\Omega = 2\pi \sin\theta\, d\theta \text{ and } U = -m\,B\cos\theta$$

After substitution, we get

$$dN = C\exp\left(\frac{mB\cos\theta}{KT}\right) 2\pi\sin\theta\, d\theta$$

Putting, $$\frac{mB}{KT} = x \qquad ...(1)$$

The component of the magnetic moment of one molecule in the direction of field is m cos θ. Therefore, the total magnetic moment of dN molecule, along the field direction is,

$$dM = dN\ m \cos\theta$$
$$= C\ \text{excosq}\ 2\pi \sin\theta\ m \cos\theta\ d\theta$$
$$= 2\pi\ C\ m \sin\theta \cos\theta\ e^{x\cos\theta}\ d\theta \qquad ...(2)$$

Since the possible orientation are contained between 1 and π, lance the total magnetic moment M of all molecules in unit volume in the direction of field obtained by integrating the above expression,

$$M = \int_0^\pi 2\pi C m \sin\theta \cos\theta\ e^{x\cos\theta} d\theta$$

$$= 2\pi\ C\ m \int_0^\pi e^{x\cos\theta} \sin\theta \cos\theta\ d\theta \qquad ...(3)$$

The total number of molecules N in unit volume can be found by integrating dN,

$$N = \int_0^\pi dN = \int_0^\pi 2\pi C e^{x\cos\theta} \sin\theta\ d\theta$$

The mean effective magnetic moment,

$$M = \frac{M}{N} = \frac{2\pi\ C m \int_0^\pi e^{x\cos\theta} \sin\theta \cos\theta\ d\theta}{2\pi C \int_0^\pi e^{x\cos\theta} \sin\theta\ d\theta}$$

$$= m \frac{\int_0^\pi e^{x\cos\theta} \sin\theta \cos\theta\ d\theta}{\int_0^\pi e^{x\cos\theta} \sin\theta\ d\theta}$$

To evaluate the integral, put cos θ = a and proceeding exactly as in the case of Langevin theory of dielectrics, we get

$$\frac{M}{m} = \frac{e^x + e^{-x}}{e^x - e^{-x}} - \frac{1}{x}$$

$$= \text{Coth}\ x - \frac{1}{x} = L\ (x) \qquad ...(4)$$

where L(x) is called "Langevin function". When x is small,

$$L(x) = \frac{x}{3} - \frac{x^3}{4 \times 5} + ...$$

Under normal experimental conditions x is very small,

$$\therefore \quad L(x) = \frac{x}{3}$$

$$\frac{\overline{M}}{m} = L(x) = \frac{mB}{2KT}, \text{ since } x = \frac{mB}{KT}$$

or $$\overline{M} = \frac{m^2 B}{3KT} \qquad \text{...(5)}$$

For paramagnetic substances,

$$B = \mu H$$

Hence, $$\overline{M} = \frac{m^2 \mu H}{3KT} \qquad \text{...(6)}$$

The volume susceptibility of the paramagnetic substance,

$$\chi_{mv} = \frac{m^2 \mu n}{3KT}, \left(\text{since} \frac{M}{H} = \chi_m\right), \qquad \text{...(7)}$$

where n are the number of molecules per unit volume.

Mass susceptibility is the susceptibility per unit mass of the substance,

$$\chi_m = \frac{n_m \mu\, m^2}{3KT} \qquad \text{...(8)}$$

where N_m are the number of molecules per unit mass of the gas.

The molecular susceptibility,

$$\chi_m = \frac{N_m \mu\, m^2}{3KT} \qquad \text{...(9)}$$

where N_m are the number of molecules in a gram molecule (Avogadro's number),

or $$\chi_m = \frac{C_m}{T} \qquad \text{...(10)}$$

where C_m is called the "Curie constant" given by,

$$C_m = \frac{N_m{}^2 \mu\, m^2}{3R}, \text{ since } K = \frac{R}{N_m} \qquad \text{...(11)}$$

χ^m is referred as "molar susceptibility" *i.e.,* susceptibility for one gram molecules of the substance.

Thus, we see that susceptibility depends on the temperature. However, the total magnetisation of a paramagnetic material must include the induced moment, and adding their contribution given in Eq. (7) to Eq. (5), we get,

$$\vec{M} = \left(\frac{m^2}{3KT} - \frac{e^2}{6m_e}\Sigma r^2\right)\vec{B} \qquad ...(11a)$$

Similarly, the total susceptibility of a paramagnetic material must include the induced susceptibility.

Failure of Langevin Theory

Langevin theory could not explain a more complicated dependence of susceptibility upon temperature exhibited by several paramagnetics such as highly compressed and cooled gases, very concentrated solutions of salt, solid salts and crystals. Langevin theory also did not explain the intimate relation observed between para and ferro-magnetism.

WEISS MOLECULAR FIELD THEORY OF PARAMAGNETISM

As stated above, the Landing theory did not explain the intimate relation observed between para and ferro-magentism. In an attempt to explain these points, Weiss in 1907, improved upon Langevin's treatment by the introduction of a new concept of internal molecular field. Weiss assumed that, since in a real gas the molecules are mutually influenced by their magnetic moments, there should exist within the gas a molecular field, produced at any point by all the molecules in the neighbourhood, proportional to and acting in the same sense as the magnetisation vector $\vec{M}$. Therefore, we have

$$\text{Molecular field } \mu \text{ Magnetisation } \vec{M}$$
$$= \gamma \vec{M}$$

where γ is a constant (molecular field coefficient). Then the effective field may be regarded as the vector sum of the external applied field $\vec{B}$ and internal molecular field *i.e.*,

$$B_{eff} = \vec{B} + \gamma\vec{M} \qquad ...(1)$$

Replacing B in equation (1) by B_{eff}, we get,

$$M = \frac{m^2(B + \gamma M)}{3KT}$$

The magnetic moment in a gram molecule,

$$M = \frac{N_m m^2(B + \gamma M)}{3KT}$$

where N_m are the number of molecules in a gram molecule. Rearranging the terms

$$M\left(1-\frac{N_m m^2\gamma}{3KT}\right)=\frac{N_m m^2 B}{3KT}$$

$$M=\frac{N_m m^2 B}{3KT\left(1-\frac{N_m m^2\gamma}{3KT}\right)}$$

$$=\frac{N_m m^2 B}{(3KT-N_m m^2\gamma)}$$

$$=\frac{N_m m^2 B}{3K\left(T-\frac{N_m m^2\gamma}{3K}\right)}$$

But $B = \mu H$

Therefore, $M=\dfrac{N_m m^2\mu H}{3K\left(T-\frac{N_m m^2\gamma}{3K}\right)}$

Molecular susceptibility,

$$\chi_m=\frac{M}{H}=\frac{N_m m^2\gamma}{3K\left(T-\frac{N_m m^2\gamma}{3K}\right)}$$

$$=\frac{N_m^{\ 2} m^2\mu}{3R\left(T-\frac{N_m^{\ 2} m^2\gamma}{3R}\right)},\ \text{since } K=\frac{R}{N_m}$$

$$C_m=\frac{N_m^{\ 2} m^2\mu}{3R}\ \text{is Curie constant}$$

and $\pi=\dfrac{N_m^{\ 2} m^2\gamma}{3R}$

then, $\chi_m=\dfrac{C_m}{T-\theta}$ ...(2)

This is the Curie Weiss law. According to this law, the susceptibility of a paramagnetic substance having molecular field varies inversely as the excess of temperature not from absolute zero, as in the case of Curie's law but from a certain critical temperature, the Curie point. The Curie point is able to explain the intimate relation observed between paramagnetic and ferromagnetic substances. Below the Curie point q, the ferromagnetism exists, but above the Curie point, the ferromagnetic properties disappear and the substance becomes paramagnetic.

DOMAIN THEORY OF FERROMAGNETISM

Magnetic effects in most substances are weak. However, a group of substances know as ferromagnetic material exhibit strong magnetic effects. The ferromagnetism is really an extreme case of paramagnetism. If the permanent dipoles, generally those resulting from electron spin, are very close together in the substance, there proves to be an effect (explained only by Quantum Theory), called "exchange", which results in a strong tendency for the spins of adjacent atoms or molecules to line up parallel to each other, even in the absence of an external field. Such a parallel orientation can extend, in an unmagnetised substance, over volumes of a considerable scale on an atomic order of magnitude (with physical dimensions of the order of 10^{-3} to 10^{-4} cm), though a small volume by ordinary standards. Such a volume is called a "domain", and an ordinary unmagnetised ferromagnetic substance consists of many domains. A strong permanent moment is associated with each domain. The domains are oriented in different directions.

In the presence of an external magnetic field, the domains change the orientation of their permanent moments, lining them up with the external fields. As the applied field becomes stronger, the moment increases, until finally with a very large external field the moment reaches a limit when all moments are parallel. This limit is called the "saturation moment". Reversing the field reverses the moments but there is an effect similar to friction, hindering this reorientation, so that by the time the external field is reduced to zero, the substance still retains a considerable moment. This is the origin of "permanent magnetism". It the external field is reversed alternately between the one direction and other, the moment lags behind the field, resulting in the phenomenon of the "hysteresis".

These properties of ferromagnetics substances are very complicated in contrast to the diamagnetic and paramagnetic substances, in which the moment is proportional to the field. The ferromangetism decreases as the temperature increases because the individual domains lose their moments at a critical temperature called the Curie temperature (about 770°C for iron). Above the Curie temperature the substance becomes paramagnetic, but the moment instead of being proportional to 1/T where T is the absolute temperature, is proportional to $1/T - \theta$, where θ is the Curie temperature, is proportional of this decrease of ferromagnetism as the temperature increases is that the tendency towards orientation of the magnetic moments which lines them up at low temperature is opposed b thermal agitation at high temperature.

Thus, the domain theory of ferromagnetism furnishes a clear picture of the phenomena that occur in a ferromagnetic material.

Table 2.1 : Three Magnetic Vectors

Name	**Symbol**	**Associated with**	**Boundary condition**
Magnetic induction or flux density	$\vec{B}$	All currents	Normal component continuous
Magnetic field strength	$\vec{H}$	True currents only	Tangential component continuous
Magnetisation (Magnetic dipole moment per unit volume)	M	Magnetisation currents only	Vanishes in a vacuum

Defining equations for $\vec{B}\,\vec{F} = q\vec{v}\times\vec{B} = I\vec{l}\times\vec{B}$

General relation among the there vectors $\vec{B} = \mu_0\vec{H} + \mu_0\vec{M}$

Ampere's law when magnetic materials are present $\oint \vec{H}.d\vec{l} = I$

(I = true currents only)

Empirical relations for certain magnetic materials $\vec{B} = \mu_r\mu_0\vec{H}$

$\vec{M} = (\mu_r - 1)\vec{H}$

COMPARISON OF STATIC ELECTRIC AND MAGNETIC FIELDS

In order to understand clearly, it is necessary to compare electric and magnetic fields and to note both their differences and their similarities.

1. *Analogy of* $\vec{B}$ *to* $\vec{E}$

By comparison, we see certain analogies. For example we, have noted in electric fields that $\vec{E}$ is defined in terms of force, while in magnetic fields $\vec{B}$ is also defined in terms of force. Hence, $\vec{B}$ may be considered as the magnetic quantity that is analogous to the electric field intensity $\vec{E}$.

2. *Analagy of* $\vec{H}$ *to* $\vec{D}$

For example in a capacitor $\vec{D}$ is directly related to the electric charge on the plates and is independent of the medium, while near a long current-carrying wire $\vec{H}$ (By $\vec{B} = \mu\vec{H}$) is directly related to the current $\left(B = \frac{\mu_0 I}{2\mu R}\right.$ in vacuum and in any medium of permeability μ, $B = \frac{\mu I}{2\pi R}$, hence $\left.H = \frac{I}{2\pi R}\right)$ and is independent of the medium. Thus, $\vec{D}$ and $\vec{H}$ may be regarded as analogous quantities.

3. *Analogy of* $\vec{H}$ *to* $\vec{E}$

However, in some other instances we may note an analogy of $\vec{H}$ to $\vec{E}$. For example, the line integral around a closed path of the total electric field intensity $\vec{E}$ is emf ($\oint \vec{E}.d\vec{l} = \xi$), while the line integral of $\vec{H}$ around a closed path is magnetomotive force (mmf = $\oint \vec{H}.d\vec{l}$). Thus $\vec{E}$ and $\vec{H}$ may be regarded as analogous quantities.

4. *Analogy of* $\vec{B}$ *to* $\vec{D}$

Furthermore in some instances we may note an analogy of $\vec{B}$ to $\vec{D}$. For example, the divergence relations,

$$\nabla.\vec{D} = r \text{ and } \nabla.\vec{B} = 0, \text{ everywhere.}$$

For charge free space (r = 0)

$$\nabla.\vec{D} = 0, \; \nabla.\vec{B} = 0$$

We note a mathematical similarity of $\vec{B}$ to $\vec{D}$.

Important Comment

If the analogy of $\vec{H}$ to $\vec{E}$ and $\vec{B}$ to $\vec{D}$ is pursued, it is possible to achieve a formal or mathematical symmetry between many of the electric and magnetic field equations. *However, electric and magnetic fields are fundamentally different, and the analogy of* $\vec{B}$ *to* $\vec{E}$ *and* $\vec{H}$ *to* $\vec{D}$ *has more meaningful physical significance. Static electric fields are due to electric charge, a scalar quantity while static magnetic fields are due to electric current, a vector quantity.* The electric and magnetic field relations developed are summarised. The analogy of $\vec{B}$ to $\vec{E}$ and $\vec{H}$ to $\vec{D}$ will be noted in many of the equations while the other analogy may be observed in a few of the relations. The relations apply to static and slowly time varying situations. They also apply in more rapidly time varying situations with the exception of those relations involving the curl or the line integral ($\vec{H} = -\nabla V_m$ also does not apply in rapidly time varying case).

Table 2.2 : A comparison of Static Electric and Magnetic Field Equations

Description of equation	Electric Fields	Magnetic Fields
Force	$\vec{F} = q\vec{E}$	$d\vec{F} = Id\vec{l} \times \vec{B}$ $\vec{F} = Q_m\vec{B}$
Basic relations for lamellar or solenoid field	$\nabla \times \vec{E}_c = 0$	$\nabla.\vec{B} = 0$
Derivation from scalar or vector potential	$E_c = -\nabla V$ $V = \frac{1}{4\pi\varepsilon_0}\int \frac{\rho}{r} dv'$	$\vec{B} = \nabla \times \vec{A}$ $\vec{A} = \frac{\mu_0}{4\pi}\int_v \frac{\vec{J}}{r} dv'$
Relation for $\vec{D}$ and $\vec{H}$	$\vec{D} = \epsilon\vec{E}$ $\nabla.\vec{D} = r_f$	$\vec{H} = \frac{\vec{B}}{\mu}$ $\nabla \times \vec{H} = \vec{J}$
Energy density	$W_e = \frac{1}{2}\epsilon E_2$	$W_m = \frac{1}{2}\frac{B^2}{\mu}$
Capacitance and inductance	$C = \frac{\text{Ch arg e}}{\text{Potential diff.}}$	$L = \frac{\text{Flux linkage}}{\text{Current}}$
Closed path of integration	$\oint \vec{E}.d\vec{l} = \text{emf}$ $\oint \vec{E}.d\vec{l} = 0$ and	$\oint \vec{H}.d\vec{l} = NI = \text{mmf}$ $\oint \vec{H}.d\vec{l} = 0$ (no current enclose)
Derivation from scalar potentials	$\vec{E}_c = -\nabla V$	$H = -\nabla V_m$ (In current free region)

*$\vec{E}_c$ is the static electric field intensity (due to charges).

$\vec{E}$ (without subscript) implies that emf producing fields (not due to charges) may also be present.

COMPARISON OF ELECTRIC AND MAGNETIC RELATIONS INVOLVING POLARISATION AND MAGNETISATION

It is interesting to compare the magnetic relations where magnetisation $\vec{M}$ is present with the corresponding electric relations where polarisation $\vec{P}$ is present. This is done in Table 2.3.

Table 2.3: Comparison of Equations involving polarisation $\vec{P}$ and Magnetisation $\vec{M}$

Description of Equation	Electric case	Magnetic case
Dipole moment relation	$\vec{P} = \frac{\vec{P}}{v} = \frac{Q\vec{d}}{v}$	$\vec{M} = \frac{\vec{m}}{v} = \frac{Q_m \vec{l}}{v}$
Flux density	$D = \left(\varepsilon_0 + \frac{\vec{P}}{\vec{E}}\right)\vec{E}$	$\vec{B} = \mu_o \left(1 + \frac{\vec{M}}{\vec{H}}\right)\vec{H}$
Permittivity and permeability	$\varepsilon = \varepsilon_0 + \frac{\vec{P}}{\vec{E}}$	$\mu = \mu_o \left(1 + \frac{\vec{M}}{\vec{H}}\right)$
Relation to polarisa tion charge density and to equivalent current dainty	$\nabla.\vec{P} = -\rho_p$	$\nabla \times \vec{M} = \vec{J}_m$
Poisson's eqn.	$\nabla 2V = -\frac{\rho}{\varepsilon_0}$ in vacuum $\nabla 2V = -\frac{\rho}{\varepsilon}$ in medium	$\nabla 2H_m = \nabla.\vec{M} = -\nabla.\vec{H}$
Scalar and vector potentials	$V = \frac{1}{4\pi\varepsilon_0}\int_v \frac{\rho - \nabla' P}{r} dv'$	$\vec{A} = \frac{\mu_o}{4\pi}\int_v \frac{\vec{J} + \nabla \times \vec{M}}{r} dv'$
Magnetisation curve	$\vec{P} - \vec{E}$ curve	$\vec{M} - \vec{H}$ curve identical to $\vec{P} - \vec{E}$ curve
Susceptibility	$\epsilon = \epsilon_0 (1 + \chi_{er})$ χ_{er} dimensionless	$\mu = \mu_o (1 + z_m)$ z_m is dimensionless

DETERMINATION OF MAGNETIC SUSCEPTIBILITY OF DIA AND PARAMAGNETIC SUBSTANCES

Let a small piece of a given material be suspended in the magnetic field of an electromagnet as shown in Fig. 2.2. Further, let the magnetic field between pole pieces be non-uniform *i.e.*, it varies from place to place. The suspended substance will experience a force. As the diamagnetism is an induced effect *i.e.*, magnetisation is in the opposite direction to the magnetising field, hence if the sample under consideration is of *diamagnetic material it will tend to move to the region of smaller magnetic field.* On the other hand, since for paramagnetic substances the magnetisation is in the same direction as the magnetising field hence if the sample is of *paramagnetic material it will move towards the region of higher magnetic field.* Suppose a force F acts on the sample placed in the non-uniform magnetic work done by this force will be F.dx. *This work done will be equal to the change in potential energy caused by the displacement of the sample.* Let a small volume v of the given materiai be situated at the place where B is magnetic field and the permeability of the medium (air) occupying the field when the sample is outside the field is $\mu\alpha$. Τηε energy density of the magnetic field on the medium (air) is given by eq. (1), *i.e.*,

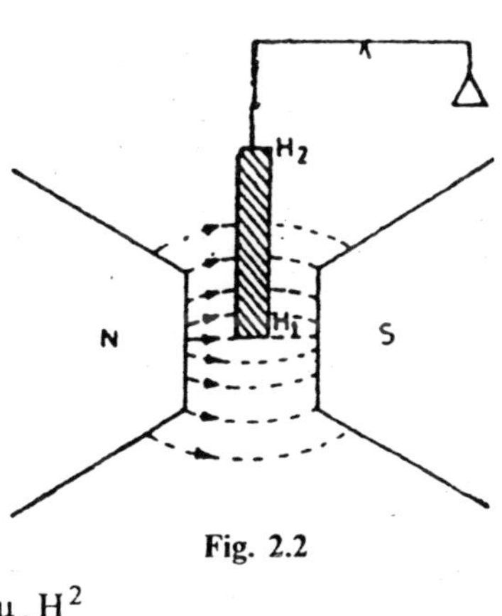

Fig. 2.2

$$\frac{B^2}{2\mu_a} = \frac{\mu_a H^2}{2}$$

Thus, the volume v which is to be occupied by the sample contains magnetic energy of amount,

$$U_1 = \frac{\mu_2 H^2}{2} v$$

Now, the sample of permeability m is placed in the magnetic field. Then the magnetic energy contained by the volume v is

$$U_2 = \frac{\mu H^2}{2} v.$$

Therefore, when a volume v of the given material replaces the same volume of air, it contains an *excess amount of energy given by,*

$$U_2 - U_1 = \frac{(\mu - \mu_a) v H^2}{2}$$

As the magnetising field H is not uniform, then the sample experiences a force and consequently moves through a distance dx. The change in the magnetic energy during this motion is,

$$\frac{d}{dx}\left[\frac{(\mu - \mu_a) v H^2}{2}\right] dx = \frac{(\mu - \mu_a)}{2} v 2H \frac{dH}{ddx} dx$$

$$= (\mu - \mu_a)\, vH \frac{dH}{dx} dx$$

$$\frac{\mu}{\mu_o} = \chi_m + 1 \text{ or } \mu = \mu_o (\chi_m + 1)$$

where χ_m is magnetic susceptibility of material. For air we have

$$\mu_a = \mu_o (\chi_{ma} + 1)$$

where χ_{ma} is the magnetic susceptibility of air. Therefore,

$$(\mu - \mu_a) = \mu_o [\chi_m - \chi_{ma}]. \text{ Hence}$$

$$(\mu - \mu_a)\, vH \frac{dH}{dx} dx = \mu_o (\chi_m - \chi_{ma})\, vH \frac{dH}{dx} dx$$

This change in magnetic energy is equal to the work done on the sample. Thus,

$$F dx = \mu_o [\chi_m - \chi_{ma}]\, vH \frac{dH}{dx} dx$$

or $$F = \chi_o [\chi_m - \chi_{ma}]\, vH \frac{dH}{dx} \quad ...(1)$$

By using this equation we can determine the susceptibility χ_m of a dia or paramagnetic material, provided magnetic susceptibility of air is known. We shall discuss two methods.

1. Gouy Method when the substance is available in the form of solids.

2. Quincke's method when the substance is available in the form of liquids.

1. Gouy Method for Measuring Susceptibility of Solids

The sample under test is taken in the form of thin rod and is suspended from one arm of a sensitive micro balance in the magnetic field of electromagnet. The field will vary rapidly along the vertical direction, hence the pole pieces are wedge shaped. If the pole pieces are sufficiently wide apart, the variation in the field in the horizontal direction will be small. Hence the field is uniform along horizontal direction.

As the vertical dimension of the pole pieces is small, the magnetic field falls rapidly in the vertical direction. Procedure is as follows.

Put the specimen at the central part of the field where the field is uniform horizontally. First the specimen is balanced by putting weights on the other arm of the balance. Now the magnetic field is switched on and the force (upward or downward depending on whether the specimen is dia or paramagnetic) acts on the specimen due to the non-uniform field in vertical direction. This force is counter balanced by adjusting the weights to balance the specimen. If the weights added or removed in the second case be m, then the force experienced by specimen due to magnetic field is mg. If A is the area of the cross-section of rod, then volume of dx length of the rod is Adx where x is measured vertically from upper end. Then the force on dx length of the rod is by eq. (1),

$$dF = \mu, [\chi_m - \chi_{ma}] \, AdxH \frac{dH}{dx}, \; [v = Adx]$$

Therefore, the force experienced by the whole rod is

$$F = \mu_o [\chi_m - \chi_{ma}] \, A \int_{H2}^{H1} HdH$$

$$F = \frac{1}{2} \mu_o (\chi_m - \chi_{ma}) \, A \, (H_1^2 - H_2^2) \qquad ...(2)$$

where H_1 and H_2 are the values of magnetic field at the lower and upper ends of the rod.

But, F = mg, hence

$$mg = \frac{1}{2} \mu_o \, A \, (\chi_m - \chi_{ma}) \, (H_1^2 - H_2^2)$$

As magnetic field falls very rapidly upward from the centre, hence H_2 is very small it can be neglected. Therefore,

$$mg \approx \frac{1}{2} \mu_o A (\chi_m - \chi_{ma}) H_1^2 \qquad ...(3)$$

If χ_{ma} is neglected as it is very small is comparison to cma then,

$$mg = \frac{1}{2} \mu_o A \chi_m H_1^2 \qquad ...(4)$$

Hence by using eq. (3) or (4) χ_m can be calculated. Magnetic field H_1 is measured by fluxmeter.

2. Quincke's Method for Measuring Susceptibility of Liquids

Quincke directly used Gouy method for liquids. The liquid under test is taken in a U tube, one end of which is capillary section and the other end as wide tube.

As shown in Fig. 2.3, the capillary section is placed between the pole pieces of electromagnet. The liquid meniscus is placed in the centre of the pole pieces where the field is uniform. When the current energising the electromagnet is switched on under the action of magnetic field the level of meniscus falls or rises depending on whether liquid is diamagnetic or paramagnetic. This change in the height of the meniscus is measured by a travelling microscope. If it is h, then h is the height of the liquid column supported by the force due to non-uniform magnetic field.

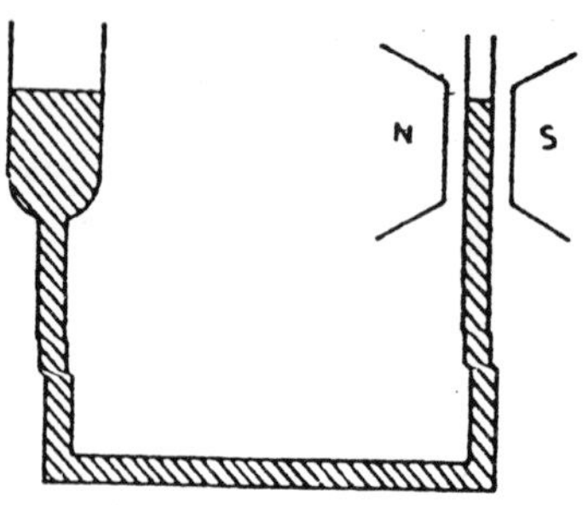

Fig. 2.3

If ρ be the density of the liquid and σ be that of air , then change in pressure of the liquid corresponding to the change h is $(\rho - \sigma)$ gh. Therefore, force acting downwards on the liquid

$$= (\rho - \sigma) \, gh \, A \qquad ...(5)$$

where A is area of cross-section of the capillary tube.

The vertical force exerted by magnetic field is given by eqn. (1) is,

$$dF = \mu_o (\chi_m - \chi_{ma}) \, Adx \, H \frac{dH}{dx}$$ where v has been replaced by Adx. The total force can be obtained by integrating the above equation from H = 0 *i.e.,* the initial zero magnetic field to $H = H_m$, the final value of magnetic field,

$$F = \mu_o [\chi_m - \chi_{ma}] \int_{H=0}^{H=H_m} Adx \frac{HdH}{dx}$$

or $$F = \frac{1}{2} \mu_o (\chi_m - \chi_{ma}) \, A \, H_m^2 \qquad ...(6)$$

Equating (5) and (6), we get,

$$(\rho - \sigma) \, gh = \frac{1}{2} \mu_o (\chi_m - \chi_{ma}) \, H_m^2 \qquad ...(7)$$

If the susceptibility of air χ_{ma} is known we can calculate χ_m from equation (7). The H_m can be measured by fluxmeter.

MAGNETIC MOMENT AND ANGULAR MOMENTUM (BOHRMAGNETON)

An electron is circulating in a circular orbit of radius r is equivalent to current loop with the current flowing in a direction opposite to v (the velocity of electron). Let us examine this. An electron is moving in a orbit of radius r with velocity v The motion of electron in orbit is governed by Bohr's rule. The force between an electron in the orbit of radius r and the nucleus of hydrogen atom is given by Coulomb's law.

$$F_e = \frac{1}{4\pi\varepsilon_o} \frac{e^2}{r^2}$$

and electron is bound by this force with nucleus, but it moves in a circular orbit of radius r. The centripetal force $Fc = \frac{m_e v^2}{r}$ holding the electron in an orbit r from the nucleus is provided by the electrostatic force $F_e = \frac{1}{4\pi\varepsilon_o} \frac{e^2}{r^2}$ between them, and the condition for orbit stability is $F_c = F_e$ *i.e.,*

$$\frac{1}{4\pi\varepsilon_o} \frac{e^2}{r^2} = \frac{m_e v^2}{r} \qquad ...(1)$$

or $$v = \sqrt{\frac{e^2}{4\pi\varepsilon_0 m_e r}}$$

The angular velocity ω is given by

$$\omega = \frac{v}{r} = \sqrt{\frac{e^2}{4\pi\varepsilon_0 m_e r^3}} \qquad ...(2)$$

The orbital angular momentum is

$$L_l = m_e \, vr$$

Substituting v from eqn. (1), we get,

$$L_l = \sqrt{\frac{e^2 m^2 r}{4\pi\varepsilon_0}} \qquad ...(3)$$

Its direction is determined by cross-product rule, as $\vec{L}_l = \vec{r} \times \vec{p}$, where $\vec{p} = m\vec{v}$. Thus L_l is perpendicular to $\vec{r}$ and $\vec{v}$, if $\vec{r}$ is along x-axis, $\vec{v}$ is along y-axis then $\vec{L}_l$ is along z-axis. $\vec{L}_l$ is directed perpendicular to the plane of the orbit as shown upward.

Now the rotating electron constitute the current, and the current is determined by the rate at which the charge passes the given point. If v is the velocity of the electron it will make $\frac{v}{2\pi r}$ revolutions per second and is equivalent to a current given by

$$i = \frac{ev}{2\pi r}$$

$$= e\frac{\omega}{2\pi}, \text{ substituting w from above Eqn. (2)}$$

$$= \frac{e}{2\pi}\sqrt{\frac{e^2}{4\pi\varepsilon_0 m_e r^3}}$$

$$= \sqrt{\frac{e^4}{16\pi^2\varepsilon_0 m_e r^3}}$$

The close orbit will behave as a magnetic dipole and its magnetic moment is given by

$$m_l = iA$$

$$= \sqrt{\frac{e^4}{16\pi^2\varepsilon_0 m_e r^3}} \cdot \pi r^2$$

$$= \frac{e^2}{4}\sqrt{\frac{r}{\pi\varepsilon_0 m_e}} \qquad ...(4)$$

Its direction is given by right hand curl rule. Let the fingers of the RH curl around the loop in the direction of the current, the extended right thumb will then point in the direction of $\vec{m}_l$. However, by relation j = nev and i = j/A the direction of current is in the direction of v.

Thus ml is also directed perpendicular to the plane of the orbit, so Ll and ml are in the same direction. The ratio

$$\frac{m_l}{L_l} = \frac{e^2}{4}\sqrt{\frac{r}{\pi\varepsilon_0 m_e}}\sqrt{\frac{e^2 m_e r}{4\pi\varepsilon_0}} \text{ (substituting from eqns. 3 and 4)}$$

$$= e/2\ me \qquad ...(5)$$

which shows that the orbital magnetic moment of an electron is proportional to its orbital angular momentum.

ORBITAL GYROMAGNETIC RATIO

The ratio of the magnetic dipole moment of the electron due to its orbital motion and the angular momentum of the orbital motion is called orbital gyromagnetic ratio, or sometimes orbital magneto mechanical ratio of an electron. Thus,

$$\frac{\text{magnetic moment}}{\text{Angular momentum}} = \frac{e}{2m_e}$$

From Bohr's postulates, the electron chooses such orbits, for which the angular momentum L_l is an integral of h/2π where h is Planck's constant *i.e.*,

$$m_l = \frac{nh}{2\pi}\frac{e}{2m_e} \qquad ...(1)$$

$$= n\left(\frac{eh}{4\pi m_e}\right)$$

where n = 1, 2, 3, ... etc. ...(2)

All the quantities in $\frac{eh}{4\pi m_e}$ are constants and together form a unit of magnetic moment called the Bohr magneton, *i.e.*,

$$\text{Bohr magneton} = \frac{eh}{4\pi m_e} \quad ...(3)$$

substituting e, h and m, we get,

$$\text{Bohr magneton} = 9.1 \times 10^{-24} \text{ amp.m}^2$$

And for first orbit n = 1, from equation (3), the magnetic moment is one Bohr magneton.

ELECTRON SPIN

S.A. Goudsmith and G.E. Uhlenbeck proposed in 1925 that an electron while moving in circular orbit round the nucleus also rotates about its own axis just as the earth rotates about its axis. In other words the electron has a spin motion over and above the orbital motion. The rotation involves angular momentum, and because the electron is negatively charged it has a magnetic moment Ms. opposite in direction to its angular momentum vector L_S as shown in Fig. 2.4.

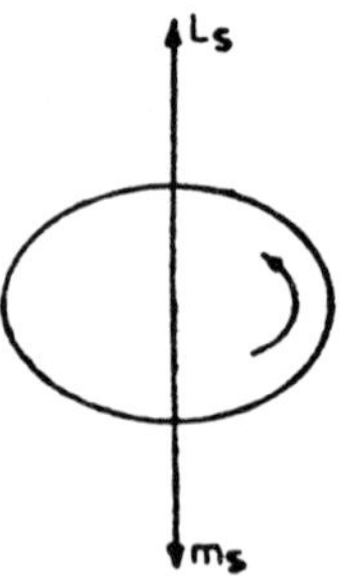

Fig. 2.4

With the idea of spin the electron possess an intrinsic angular momentum independent of any orbital angular momentum it might have and associated with this angular momentum is a certain magnetic moment m_S. The detailed theory which is purely quantum mechanical is beyond the scope here, however, the ratio of magnetic moment m_S to the angular momentum L_S for the electron spin is twice as that for its orbital motion, *i.e.*,

$$\frac{m_S}{L_S} = \frac{e}{m_e}, \text{ [from equation (1)]} \quad ...(1)$$

The magnitude of L_S always comes out to be $\frac{1}{2}\left(\frac{h}{2\pi}\right)$ where h is Planck's constant.

In an atom there are several electrons each one has angular momentum as well as magnetic moment for both orbital and spin motion. Thus, an atom have a total angular momentum and a total magnetic moment. For an atom as a whole the ratio of magnetic moment to angular momentum on the basis of quantum physics is e/m, or e/2m, or in between these two because there is a combination of the contributions from the orbits and spins,

$$m_A = g\left(\frac{e}{2m_e}\right)L \qquad ...(2)$$

where m_A is total magnetic moment and L is total angular momentum and g is called Lande g-factor which is the characteristic of the atom. The value of g is one for a pure orbital moment, two for a pure spin moment and for an atom g lies between one and two because electrons in atoms have both orbital and spin motion.

In nuclei there are protons and neutrons and they also more around in orbits and spin like electrons. The magnetic moment is parallel to the angular momentum. Therefore for nuclei, we have,

$$m_{nuclei} = g\left(\frac{q_e}{2m_p}\right)L \qquad ...(3)$$

where q_e is charge, m_P is mass of proton and g is called the nuclear g factor, q is a number near one to be determined for each nucleus. However the spin magnetic moment of the proton does not have a g factor of two, *i.e.*, the value for the electron. For proton g = 2.79. The neutron although neutral also has a spin magnetic moment, The neutron is not exactly neutron in the magnetic sense.

ENERGY LOSS PER CYCLE OF MAGNETISATION

The process of magnetisation of a ferromagnetic substance through a cycle is involves the expenditure of the energy. This energy is spent in aligning the elementary dipoles in the direction of magnetising field. If we gradually reduce the magnetising field to zero, the magnetisation does not fall to zero and magnetic field has to be applied in the opposite direction to completely demagnetise the sample. Thus energy spent in magnetising a specimen is not completely recoverable and there is a loss of energy in taking a sample through a cycle of magnetisation and it appears in the form of heat. This loss of energy is called hysteresis loss.

Hysteresis Loss from B-H Curve

Let us consider the sample of ferromagnetic material in the shape of toroidal solenoid. The magnetising field H is obtained by modified form of Ampere's law and is given in equation (1) *i.e.*, $\oint \vec{H}.d\vec{l} = I$, $H_l = NI$ or $H = \frac{NI}{l}$ where N are total number of turns over toroidal winding and l is length of toroid

thus $H = n_l I$

where nl are the number of turns per unit length, then we get,

$$\vec{H} = \frac{NI}{l} \quad ...(1)$$

Let us calculate the amount of work that has to be done in establishing a current I in the toroidal coil by which magnetising field H is obtained and the substance is magnetised giving rise to the magnetic flux density (or induction) $\vec{B}$. If A is the area of toroidal ring, the flux inside the ring is,

$$\phi = NAB$$

Now as current I increases, the magnetic flux also increases and according to Faraday's law an induced emf develops in the coil, the magnitude is given by,

$$\xi = \frac{d\phi}{dt} = NA\frac{dB}{dt}$$

The rate of doing work,

$$P = \xi I = NAI.\frac{dB}{dt}$$

Substituting I from equation (2),

$$P = HA_l\frac{dB}{dt}$$

The product A_l is the volume of the sample *i.e.*, ring. Hence the rate of doing work per unit volume,

$$P = H\frac{dB}{dt}$$

This gives the rate at which energy is transferred per unit volumes to the sample in magnetising it is,

$$\frac{dw}{dt} = P = H\frac{dB}{dt}$$

Therefore the energy transferred in time dt per unit volume,

$$dw = HdB$$

The energy transferred to the sample per unit volume per cycle

$$= \oint HdB \qquad \text{...(3)}$$

Now be careful to the fact that through it has been deduced for an iron ring this expression is true in general.

Let us evaluate the integral for ferromagnetic materials. Ferromagnetic materials are "inherently irreversible" as they retain magnetisation even if the magnetising field is reduced to zero. Fir. 2.5 represents a part of B-H loop. For ferromagnetic materials the small change in H produces a large change in $\vec{B}$ (or $\vec{M}$). But for the change dH in H, the corresponding change dB in B will be very small. Therefore, the value of H may be taken to be same for the points a and b. The product HdB is the energy transferred per unit volume to the sample for a change dB in B is equal to the area ab × cb which is area abcd shown shaded. Therefore, the energy transferred for the path OA per unit volume

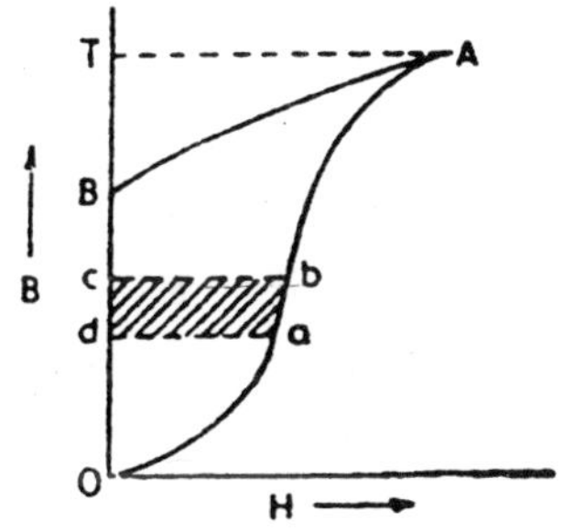

Fig. 2.5

$$= \int_{OA} HdB$$

$$= \text{area OATO}$$

Now H is gradually decreased reduced to zero as the process is irreversible. The trace does not follow AO but AB. Hence, the energy recoverable during demagnetisation

$$= \text{area BATB}$$

Therefore the energy Energy transferred –Energy recovered

not recoverable = during magnetisa-during demagneti-

i.e., energy losstion sation

= area OATO -- area BATB

= area OABO

If we consider the hysteresis loop, then the energy not recoverable, *i.e.*, energy loss,

$$\oint HdB = \text{area ABCDEFA}$$

$$= \text{area B-H loop} \qquad ...(4)$$

The energy loss is converted into heat.

Energy density

The energy required to magnetise the sample per unit volume per cycle is given by eqn. (4), *i.e.*,

$$= \oint HdB$$

It is assumed that the relation $B = \mu H$ is valid over the entire range of B with $\mu = \mu_o \mu_r$ then energy stored in the sample if B changes from 0 to B.

$$\int_0^B HdB = \int_0^B \frac{B}{\mu} dB = \frac{B^2}{2\mu} = \frac{\mu H^2}{2} = \frac{BH}{2} \text{ Joules/metre}_2$$

Magnetic energy per unit volume is called,

Magnetic energy density

$$= \frac{1}{2}\frac{B^2}{\mu} \text{ Joules/metre}^2$$

$$= \frac{1}{2}\mu\, H_2 \text{ Joules/meter}^2 \qquad ...(5)$$

$$= \frac{1}{2} BH \text{ Joules/metre}^2$$

where the permeability of the medium is μ. In case the medium is vacuum (or air), $\mu_r = 1$, $\mu = \mu_o$.

$$\text{Energy density} = \frac{1}{2}\frac{B^2}{\mu_o}$$

which is same as derived earlier in equation (5). This expression for energy is analogous to the electrostatic energy expression *i.e.,*

$$\frac{1}{2} \epsilon_0 E^2 \text{ or } \frac{1}{2} \varepsilon E^2$$

Hysteresis Loss from MH. Cycle

Suppose the behaviour of the ferromagnetic substance is represented by M-H curve. Using the relation,

$$\vec{B} = \mu_0 (\vec{H} + \vec{M})$$

or $$dB = \mu_0 (dH + dM)$$

Integrating for whole hysteresis loop, we have after multiplying by H,

$$\oint HdB = \mu_0 \oint HdH + \mu_0 \oint HdM$$

If H is plotted against dH we will get a straight line, the integral $\oint HdH$ *represents the area enclosed by a straight line which is zero.* So we have,

$$\oint HdB = \mu_0 \oint HdM$$

$\oint HdB$ represents the energy loss per unit volume per cycle and equal to area of $\vec{B}$-$\vec{H}$ loop. Similarly $\oint HdM$ equals to area of $\vec{M}$-$\vec{H}$ loop or curve.

Therefore the energy loss = μ_0 area of M-H loop ...(6)

This is called Warburg's law.

Thus, *the energy loss per unit volume of the material per cycle is equal to* μ_0 times the area of M-H loop and is dissipated in the form of heat.

DEMAGNETISATION

Ferromagnetic materials are not fully demagnetised by removing the magnetising field. They retain magnetisation corresponding to H = 0. The capacity to retain magnetisation depends on the degree up to which the substance is magnetised. It is maximum corresponding to saturation value of magnetisation. The portion of a saturation hysteresis loop for which M is positive H is negative is called "demagnetisation curve". The retentivity is different for different values of H below saturation value.

In fact retentivity decreases with decreasing H below saturation value. Hence it is possible to reduce retentivity to zero value by gradually decreasing H, through repeated cycles as shown in Fig. 2.6.

However the demagnetisation is also possible by heating the substance above the Curie temperature. But this method causes the permanent change in the magnetic behaviour of the substance.

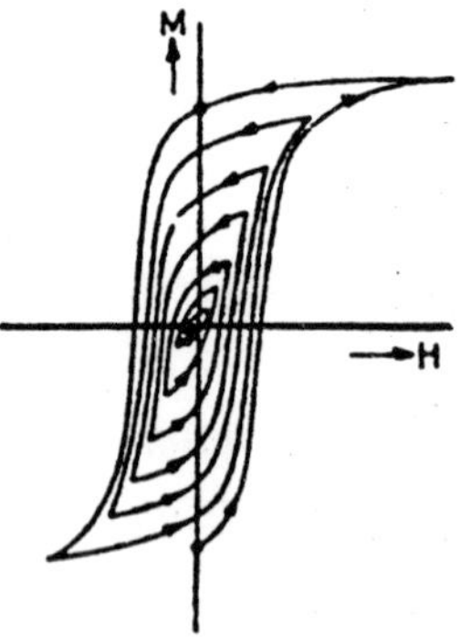

Fig. 2.6

Soft and Hard Magnetic Materials

The shape of hysteresis loop is a characteristics of the magnetic material under test. Fig. 2.7 shows M-H curves for iron and steel. We can draw the following conclusions from the curves.

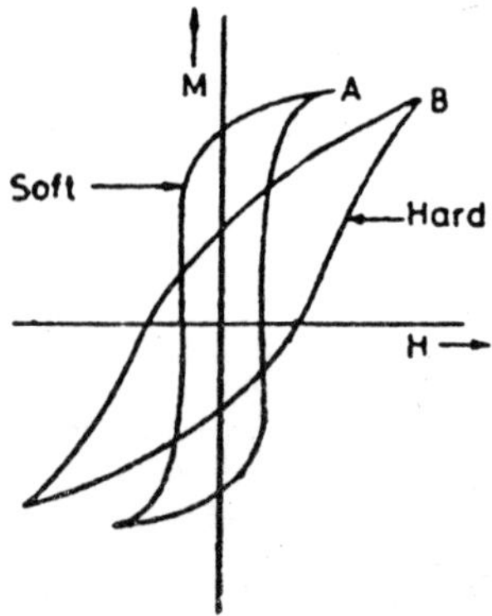

Fig. 2.7

Coercivity

Substances with large value of coercive force are said to be magnetically hard, whereas those small coercive force are called soft.

Retentivity

The retentivity is greater for soft materials than for hard material.

Susceptibility

It is greater for soft materials than for hard materials.

Permeability

It is greater for soft materials than for hard materials.

Hysteresis Loss

The area of the hysteresis loop which measures the energy loss per unit volume per cycle is less for soft materials than for hard materials.

USES OF HYSTERESIS CURVE CHOICE OF MATERIALS

The hysteresis curve and its properties, saturation, retentivity, coercivity and hysteresis loss provide very useful information for choosing the material for practical uses in industry

Permanent Magnets

For permanent magnets the material should have

(1) Large coercive force because we want that its magnetism should be permanent and not easily destroyed by magnetic fields, mechanical ill-treatment or by temperature change, and

(2) High residual magnetism so that the magnet may be strong.

These requirements are satisfied by steel. Though the retentivity of iron is greater than steel but iron has very small value of coercivity, *hence steel is preferred for permanent magnets.* The fact that hysteresis loss is also less for soft iron is immaterial because permanent magnets are never put to cyclic change of magnetisation. The materials used are cobalt-steel, tungsten steel and other number of alloys. Very recently an alloy of iron, cobalt and vanadium called vicalloy having highest coercivity has been developed.

Electromagnets

The materials for the construction of core of electromagnets should have

1. maximum induction B with comparatively small magnetising field H.
2. high initial permeability.
3. low hysteresis loss.

Considering these facts the *soft iron is an ideal material for constructing an electromagnet.*

Transformer Cores, telephone diaphragms, Armature of dynamos and motors and cores of chokes

When these are used practically the material is subjected to cyclic changes. The chief requirements are that the material must have,

1. high initial permeability in order to obtain large B for low value of H since ($B = \mu H$).
2. low hysteresis loss which will produce a small heating effect hence winding is secure of any breakdown of insulation, and
3. high specific resistance to reduce eddy current losses.

Soft iron is better than steel for these purposes. It has initial permeability about 250, initial permeability can be increased by adding 4 per cent silicon and alloy is called transformer steel which is a very good material for cores of transformers. There are some other alloys containing nickel iron called Term alloys, Mumetal (Ni 76%, Cu 5%, Fe 17.5%, Cr 1.5%) having initial permeability between 10,000 and 100,000. An alloy called radio-metal (N 48%, Fe 48.5%, Cu 3%, Mn 0.5%) has an initial permeability between 2,000 to 15,000.

Magnetomotive Force and Magnetic Circuits

We are familiar with an electrical circuit and Ohm's law which connects the emf with the current in the circuit. There is flow of electrons which constitutes the electric current. The line integral of electric field $\vec{E}$ *i.e.*, $\oint \vec{E} . d\vec{l}$ over a circuital path is the electromotive force and is equal to the work done in carrying a unit positive charge once round the circuit. But in magnetism, since there is nothing like a separate north pole and a thing as magnetic currents is meaningless. However, the magnetic lines of flux form closed loops. Such closed paths are known as magnetic circuit and we have Ampere circuital law that gives the line integral of magnetic field H *i.e.,*

$$\oint \vec{H}.d\vec{l} = \text{enclosed current}$$

Hence by analogy we can say that the line integral $\oint \vec{H}.d\vec{l}$ *is equal to the work done in currying unit north pole once round the close path if north pole is free to do so and termed as mmf (magnetomotive force)*

Hence, from Ampere's law,

$$\oint \vec{H}.d\vec{l} = H\oint dl = I$$

Here $\vec{H}.d\vec{l} = Hdl$ as H is everywhere parallel to the path $d\vec{l}$ and $\vec{H}.\oint dl$ = length of the path.

Let us consider the case of toroidal solenoid of N turns. The total current enclosed is NI, the magnetic field H inside is given by,

$$H = \frac{NI}{l} \text{ amp/metre}$$

The flux density inside the right is,

$$B = \mu H$$

$$= \frac{\mu HI}{l}$$

Total magnetic flux $\phi_B = \int_S \vec{B}.d\vec{S}$

$$\phi_B = BA = \frac{\mu NIA}{l}$$

where A is area of cross-section of ring. Therefore,

$$NI = \frac{1}{\mu A}\phi_B \qquad ...(1)$$

Returning now to the consideration of electric field, we know that emf,

$$\xi = \oint \vec{E}.d\vec{l}$$

In a magnetic field we may write an "analogous relation", based on Ampere's law, that when current is enclosed by the path of integration,

$$\oint \vec{H}.d\vec{l} = NI = \xi_m \qquad ...(2)$$

where the quantity ξ_m, called "magneto motance", or magnetomotive force" *i.e.,* mmf is equal to the current enclosed. The product NI is expressed in ampere turns.

Unit of mmf

From above equation (2), the mmf has the same unit as NI *i.e.*, ampere turns. In N = 1, then simply ampere. Comparing equation (2.1) and (2),

$$\text{mmf} = NI = \phi_B \frac{1}{\mu A} \qquad ...(3)$$

In an electrical circuit we have similar relation, namely.

$$\text{emf} = \text{current} \times \text{resistance}$$

(by Kirchhoff's law the total emf in the circuit is equal to the total IR drop). We can write,

$$\text{Current} = \frac{\text{emf}}{\text{Resistance}}, \text{ i.e., } I = \frac{\xi}{R} \qquad ...(4)$$

and from equation (2.59),

$$\text{Flux} = \frac{\text{mmf}}{1/\mu A}, \text{ i.e., } \phi_B = \frac{\xi m}{R_e} \qquad ...(5)$$

By analogy we define a quantity called the "reluctance" R_e which is considered to oppose the setting up of magnetic flux in magnetic circuit as resistance opposes the setting up electric current. R_e is expressed in amp. turns/weber which is reciprocal of inductance.

Therefore the unit for reluctance are:

$$\frac{\text{amp. turns}}{\text{Weber}} = \frac{1}{\text{Henries}}$$

Let us consider the reluctance of portion of a magnetic circuit, since in an electric circuit we have Ohm's law,

$$R = \frac{V}{I}$$

where V is potential difference between the points.

In the analogous magnetic case reluctance R_e between two points,

$$R_e = \frac{V_m}{\phi_B} \qquad ...(6)$$

where V_m = magnetic potential difference between the points, and

$$V_m = \int_1^2 \vec{H}.d\vec{l}, \text{ just as } V = \int_1^2 \vec{E}.d\vec{l}$$

Thus, we call equation (6) as "*Ohm's law*" *in magnetism.*

Permeance

The reciprocal of reluctance is known as permeance

$$\frac{1}{R_e} = \text{Permeance, its unit is } \frac{\text{Weber}}{\text{amp.}} = \text{henrys}$$

and ϕ_B = mmf × permeance

$$\text{Reluctance} = \frac{l}{\mu A}$$

and Resistance $= r\,\frac{l}{A} = \frac{1}{\sigma}\frac{l}{A}$

where σ is electrical conductivity, μ plays the same part in magnetic circuit as σ does in electrical circuit.

Combination of reluctances

If a magnetic circuit consists of different materials, with different length and cross-section. The combined reluctance is obtained by adding individual reluctances, as we do for resistances in series. Thus in a magnetic circuit made up of different materials of permeabilities μ_1, μ_2... of lengths l_1, l_2... and of cross-section A_1, A_2..., the total reluctance is given by

$$R_e = \frac{l_1}{\mu_1 A_1} + \frac{l_2}{\mu_2 A_2}$$

or $$R_e = \Sigma \frac{l}{\mu A} \qquad ...(7)$$

Points of difference in two circuits

In spite of the above analogy in two circuits with regard to various terms, they differ on the following points:–

1. In an electric circuit, the current represents the actual flow of electrons or ions whereas in magnetic circuit there is no such flow because free poles do not exist.
2. In an electric circuit, energy is required to create as well as maintain the electric current. But in magnetic circuit the energy is required only for creation of the flux.

3. Though apparently μ corresponds to σ, the specific conductivity but actually μ in magnetic process corresponds to ε which is the absolute permittivity of the medium in electrical process.
4. An important point of difference is, that the resistance is independent of the current in the circuit. As μ is not constant, the reluctance of magnetic circuit depends upon the flux.

Magnetic Circuit with Air Gap

Consider an iron ring of permeability μ, having mean circumference $l(= 2\pi r)$ and area of cross-section A and a narrow air gap of thickness d as shown in Fig. 2.16. The field H_g in the gap, $H_g = \frac{B}{\mu_o}$ while the field H_i in the iron, $H_i = \frac{B}{\mu}$

$$H_i = \frac{B}{\mu} = \frac{B}{\mu_r \mu_o} = \frac{H_g}{\mu_r}$$

or
$$\frac{H_g}{H_i} = \mu_r \qquad ...(8)$$

The reluctance of air gap $= \frac{d}{\mu_o A}$

The reluctance of iron $= \frac{l-d}{\mu A}$

Total reluctance $= \frac{d}{\mu_o A} + \frac{l-d}{\mu A}$

The magnetic flux,

$$\phi_B = \frac{NI}{\frac{d}{\mu_o A} + \frac{l-d}{\mu A}} \qquad ...(9)$$

In the absence of air gap,

$$\phi_B = \frac{NI}{l/\mu A} \qquad ...(10)$$

The comparison of equations (8) and (10) shows that the magnetic flux in the circuit with air gap is less than that in the circuit without any air gap. *Thus with air gap flux is considerably reduced.*

Magnetic Circuit of an Electromagnet

Fig. 2.8 shows an electromagnet with an air gap and with different ferromagnetic materials of various shapes and lengths. With the idea of magnetic circuit it is easy to calculate the flux by knowing the total reluctance of the circuit. The total reluctance is given by,

$$\frac{l_1}{\mu_1 A_1} + \frac{2l_2}{\mu_2 A_2} + \frac{l_3}{\mu_3 A_3}$$

where A_1, A_2 and A_3 are areas of cross-section for three parts of the electromagnet.

$$\text{Flux } \phi_B = \frac{\text{mmf}}{\dfrac{l_1}{\mu_1 A_1} + \dfrac{2l_2}{\mu_2 A_2} + \dfrac{l_3}{\mu_3 A}} = \frac{\text{mmf}}{\Sigma \dfrac{l}{\mu A}} \quad ...(11)$$

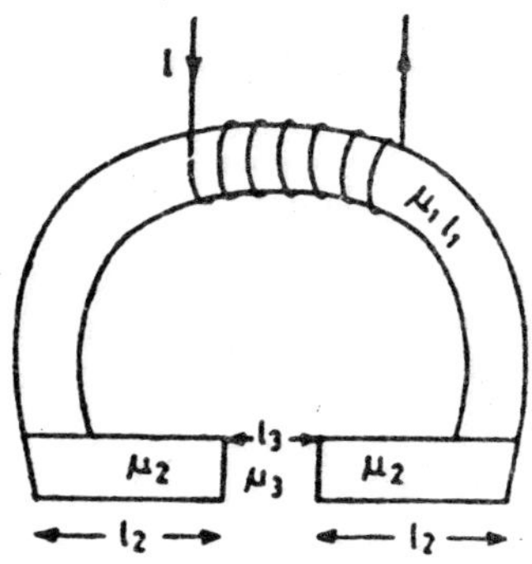

Fig. 2.8

As μ_2 corresponds to air gap $\mu_3 = \mu_o$, because μ_r the relative permeability for air is one. However, the calculation is only an approximation because leakage will take place in air gap. If air gap is very thin leakage is negligible and calculations are more reliable. *The small air gap l_3 increases the reluctance of the circuit by a large amount.*

Calculation of Lux due to a Close Ring

Such a ring is shown in Fig. 2.9. With different areas of cross-sections, A_1, A_2, A_3, ... at different points the corresponding flux densities are B_1, B_2, B_3. Assuming that there is no flux leakage and using Gauss's law for magnetism, *i.e.*,

$$\oint_S \vec{B} . d\vec{S} = 0$$

i.e., the number of lines of $\vec{B}$ leaving any volume enclosed by a surface will be the same as that entering it. Hence, we get,

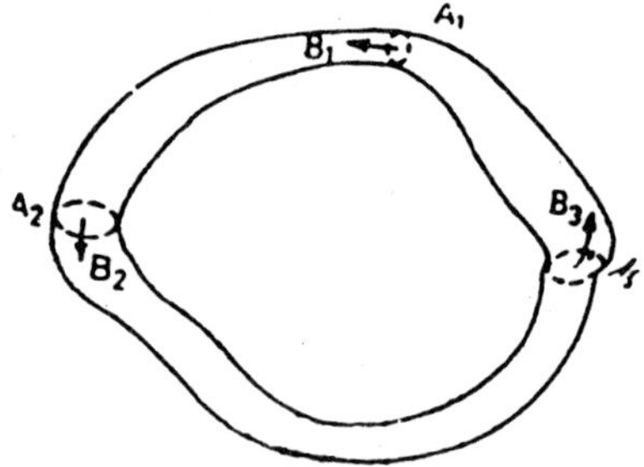

Fig. 2.9

$$\phi_B = B_1A_1 = B_2A_2 = B_3A_3 \ldots = B_nA_n$$

THE ATOMIC MODEL AND MAGNETISM

An electron revolving in its orbit around the nucleus of an atom forms a tiny electric current loop. Since a current loop has a magnetic field and *all atoms have revolving electrons we might suppose that all substances would exhibit magnetic effects.* It is reasonable to imagine that magnetic field associated with a piece of iron is due to tiny electric currents within the iron itself. Such electric currents are known as Amperian currents. It can be shown that "magnetic moment" of a current loop is m = Ia where a is the area of the orbit. This is often referred to as the "orbital magnetic moment".

Another feature of the atomic model, which is based on experimental and theoretical results is that the election in the atom also rotates about an axis through the electron (for detail see art. 4.5 electron spin). The electron spin is considered as the equivalent of a tiny current (amperian current) consequently there is an associated magnetic moment with the electron spin. This is referred to as the "spin magnetic moment". The resultant magnetic moment of the atom is obtained by the proper vector summation of the orbital and spin magnetic moments. These are combined according to quantum theory rules, which will not be discussed here. It is important to note, however, that *the magnetic properties of bulk matter can be described in terms of the orbital and spin motions of the atoms which make up the matter.*

Classification of Magnetic Material

In many atoms the electrons are completely paired, that its, for each electron spinning in one direction, there is an electron spinning in the opposite direction. And the same situation exists in regard to the orbital motion of the electrons. Thus, the net current circulating about any given axis is zero and there are no Amperian current in such substances, consequently the resultant magnetic moment of the atom is also zero. *Such substances show very "weak magnetic effects" that they are called "non-magnetic". However, vacuum is the only truly non-magnetic medium.*

All materials show some magnetic defects with the exception of the ferromagnetic group these effects are weak. It has been established both theoretically and experimentally that all matter may be classified into three groups, according to its fundamentally different behaviour under the action of a magnetic field. These are classified as (1) diamagnetic, (2) paramagnetic and (3) ferromagnetic

1. Diamagnetic Substances

Diamagnetic substances are those which, when placed in a magnetic field, *become weakly magnetised. The resultant magnetisation is opposed to the applied field.* The magnetism which is thus established, in general, varies directly with the impressed field and is independent of the absolute temperature. A diamagnetic substance contains no dipoles, but only dipoles that are induced by an external field.

2. Paramagnetic Substances

Paramagnetic substances are those which, when placed in a magnetic field. *Become weakly magnetised in the direction of the applied field.* The magnetism thus established, in general, is proportional to the impressed field and varies inversely with the absolute temperature. A paramagnetic substance contains permanent dipoles.

3. Ferromagnetic Substances

Ferromagnetic substances are those which, when placed in a magnetic field, *become strongly magnetised in the direction of the applied field,* and which also exhibit the phenomena of retentivity and hysteresis. The magnetism whitish thus established in general, is proportional to the applied field and varies with the quantity $T - \theta$, where T is the absolute temperature and θ is the critical temperature and is characteristics of the material. The variation is known as Curie Weiss law. *Ferromagnetic materials are sometimes classed as strongly paramagnetic.* All ferromagnetic substances are paramagnetic substances but all paramagnetic substances are not ferromagnetic substances. The ferromagnetism disappears are not ferromagnetic substances. The ferromagnetism disappears if the temperature is increased above a critical value called Curie temperature and substance becomes paramagnetic.

MAGNETISATION

When a magnetic material is placed in a magnetic field, the atomic currents due to uncompensated spinning electrons are aligned and the process is called magnetisation. To be more specific, since circulating electron is equivalent to a circular coil which in turn is equivalent to a small magnet. Thus the alignment of atomic currents is same as alignment of small magnets because, it is sufficient to describe them by their magnetic moment which can be expressed either as $Q_m l$ or IA, (For detail see equivalence of circular coil to a magnet). *For simplicity we define the magnetisation as the alignment of elementary (or atomic) magnetic dipoles.*

Uniform Magnetisation

Magnetisation of any substance is uniform if all the atomic magnetic dipoles are uniformly distributed throughout the volume of the substance and are oriented in the same direction as the magnetising field. Thus, magnetisation is constant throughout the material. Magnetisation is uniform if the substance is *homogeneous and isotropic.*

Magnetisation Vector $\vec{M}$

If a specimen is placed in a magnetic field then by the process of magnetisation the atoms of molecules are magnetied on the atomic scale. Precisely, the magnetisation is the interaction of magnetic dipole moments with the field in which the materials are placed. This interaction (*i.e.*, the effect of magnetising field on the magnetic dipoles can conveniently be described by a quantity called magnetisation vector $\vec{M}$. The microscopic magnetisation can very from atom to atom but can be averaged over volume which contain very many atoms to give a function which is smoothly varying with position on a macroscopic scale. This function is called the magnetisation of the medium and is represented by symbol $\vec{M}$. It is defined as the magnetic dipole moment per unit volume, and is given by

$$\vec{M} = \frac{\text{Magnetic dipole moment}}{\text{volume}} \qquad ...(1)$$

Magnetisation at a point

The value of $\vec{M}$ in above equation (2) is an average for the volume. If magnetisation is uniform, the substance has a continuous distribution of infinitesimal atomic magnetic dipoles and the dipoles are of discrete, finite size. Then, for continuous magnetisation, the value of $\vec{M}$ at a point can be defined as the net dipole moment $\sum \vec{m}$ of a small volume Δv divided by the volume with the limit taken as Δv shrinks to zero around the point, thus,

$$\vec{M} = \lim_{\nabla v \to 0} \frac{\vec{\nabla}_m}{\Delta v} \quad ...(2)$$

If the substance is not magnetised, the individual moments $\vec{m}$ will be oriented at random and the vector sum $\sum \vec{m}$ will be practically zero. In this case $\vec{M}$ is zero.

Units of $\vec{M}$.

Dimensionally eqn. (3) has the form,

$$\text{Magnetisation} = \frac{\text{Magnetic dipole moment}}{\text{volume}}$$

$$= \frac{\text{Current} \times \text{Area}}{\text{volume}}$$

$$= \frac{\text{Amp. met}^2}{\text{met}^2} = \frac{\text{Amp.}}{\text{met.}}$$

Thus, the RMKS units of magnetisation is ampere per metre.

The currents equivalent to a magnetised body

In this section we will discuss how a magnetised body can be replaced by currents in order to determine the magnetic field arising from the magnetisation. Let us first estimate the equivalent current in a simple example *i.e.*, a bar magnet.

Consider a bar magnet in which the magnetisation is every where in the direction of the positive z-axis. Let us consider the volume Δv to be a small rectangular box as shown in Fig. 2.10. Since all the atomic currents within the box may be imagined to flow around the area $\Delta x\ \Delta y$, a resultant current (or fictitious current also amperian current) I_m may be considered to flow in one direction over a face. The magnetisation $\vec{M}$ is given at a point. By,

$$M = \frac{I_m \, \Delta x \, \Delta y}{\Delta x \, \Delta y \, \Delta z} = \frac{I_m}{\Delta z} = I' \qquad ...(4)$$

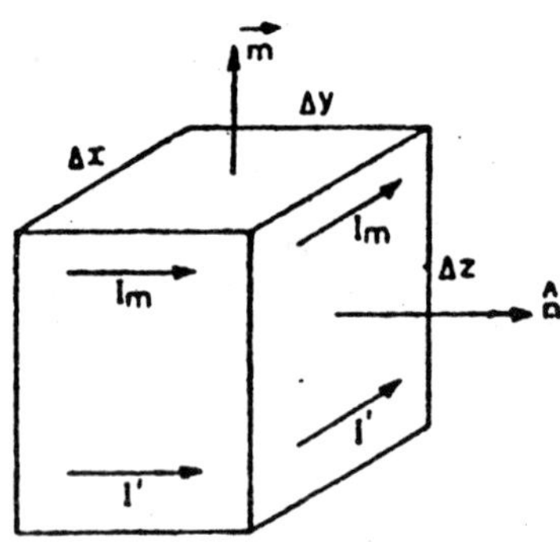

Fig. 2.10

Since magnetic dipole moment = current × area

where I' is the current flowing per unit length over the sides of the volume. Thus we must assume a surface current (I') at the surface of the magnetised volume numerically equal to the magnetisation. We can indicate the direction as well as the magnitude of this surface current by the vector equation,

$$\text{Surface current } I' = \frac{I_m}{\Delta z} = \vec{M} \times \hat{n} \qquad ...(5)$$

where $\hat{n}$ is the out-drawn normal to the surface. We call I' also as the *linear current density.*

Non-uniform Magnetisation

In this case all the atomic magnetic dipoles are distributed in a non-uniform manner. *The magnetisation varies from point to point in the material.* $\vec{M}$ is a function of position in a non-uniformly magnetised material. Then the total magnetic moment,

$$\vec{\Sigma} m = \int_V \vec{m} \, dv \text{ amp. metre}^2 \qquad ...(6)$$

where integration is carried out over the volume of the substance. Non-uniform magnetisation occurs when substances are *inhemogeneous and non-isotropic.* If at two points magnetisation is $\vec{M}$ and $\vec{M} + \Delta\vec{M}$. then

$$\text{Surface current} = \Delta\vec{M} \times \hat{n} \quad ...(7)$$

where $\hat{n}$ is normal to the surface.

If the magnetisation is inhomogenous *i.e.*, magnetisation varies from point to point in the material. There will not be the complete cancellation of the circulating currents between the boundaries of the magnetised material and there will be a net current flowing throughout the volume of the material.

Let us find the resultant current density. Consider two adjacent rectangular current loops 1 and 2 in a magnetised material. Since magnetisation is not uniform, it is different for the two loops. If $\vec{M}$ is magnetisation vector, *i.e.*, magnetic moment per unit volume. The moment $(\vec{m})$ of a single loop in Fig. 2.11 is given by,

$$\vec{m} = \vec{M}\, dx\, dy\, dz$$

where dx dy dz is the volume of the loop.

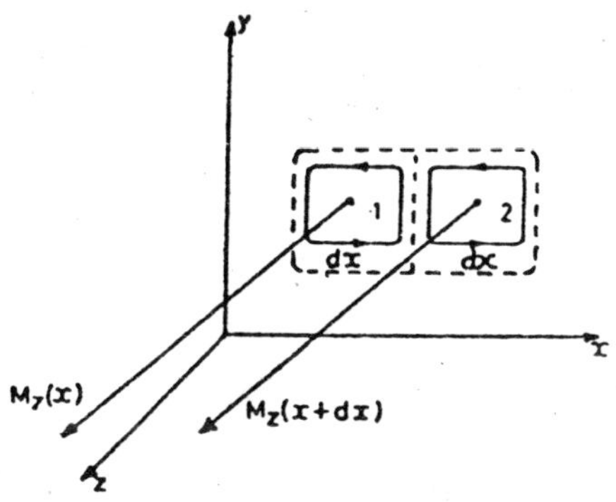

Fig. 2.11

Suppose rectangle has a magnetisation Mz in the z-direction. But we know that

$$\text{Magnetic moment} = \text{current} \times \text{area (IA)}$$

Therefore, the current (amperian or magnetising) in the rectangle 1 for the considered case if denoted by I_{m1} is given by,

$$I_{m1} = \frac{M_z dx\, dy\, dz}{dx\, dy}$$

where dx dy is the area of the coil.

If magnetisation is changing at the rate of $\frac{\delta M_z}{\delta x}$ along x-axis then magnetisation at a point where second rectangle is situated is

$$= M_z + \left(\frac{\delta M_z}{\delta x}\right)dx$$

Therefore the current in the rectangle two is

$$I_{m2} = \left[M_z + \left(\frac{\delta M_z}{\delta x}\right)dx\right]\frac{dx\ dy\ dz}{dx\ dy}$$

The difference $I_{m1} - I_{m2}$ is the net current at the mutual boundary of rectangles 1 and 2. If it is denoted by Im then,

$$I_m = I_{m1} - I_{m2} = -\frac{\delta M_z}{\delta x}dx\ dz$$

the direction is determined by equation (8) by vector product, rule Im is normal to the plane of $\vec{M}$ and $\hat{n}$ *i.e.,* in the y-direction.

However, if $\vec{M}_x$ *i.e.,* x component of $\vec{M}$ varied with z, then we should also have a component of Im in y-direction, *i.e.,*

$$= \frac{\delta M_x}{\delta z}dx\ dx$$

Thus combining these two we have the resultant current in y-direction,

$$I_{my} = \left[\frac{\delta M_x}{\delta x} - \frac{\delta M_z}{\delta x}\right]dx\ dz$$

dx dz is the area of coil in x-z plane. Therefore the current density in y direction,

$$J_{my} = \frac{I_{my}}{dx\ dz} = \left[\frac{\delta M_x}{\delta z} - \frac{\delta M_z}{\delta x}\right]$$

$$= (\text{curl } M)y$$

We have similar expressions for x and z components *i.e.,*

$$J_{mx} = (\text{curl } M)_x$$

and $$J_{mz} = (\text{curl } M)_z$$

Hence, we can generalise,

$$\vec{J}_m = \text{curl } \vec{M} \qquad ...(9)$$

Important Points

1. In fact, the amperian currents (or atomic currents) are fictitious and are replaced by the effective magnetising currents that give the same magnetic effect of the material, we shall discuss this in the next article.

2. The magnetisation $\vec{M}$ is analogous to the electric polarisation $\vec{P}$. *The amperian current in magnetic materials corresponds to the induced charges in the dielectrics.*

Three Magnetic Vectors

If we place a magnetic material in a magnetic field the elementary magnetic dipoles will be aligned and set up their own field, which will modify the original field.

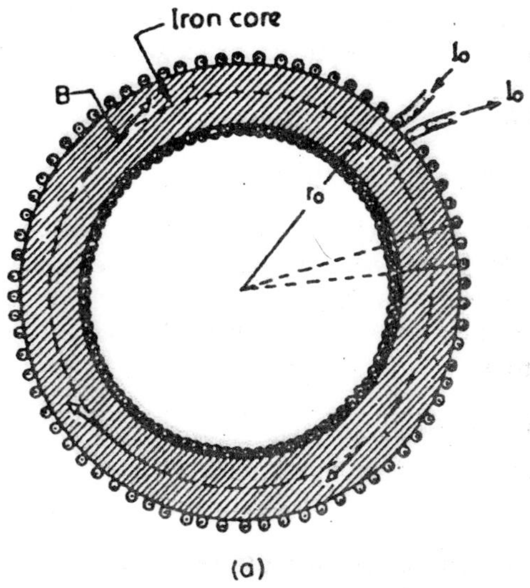

Fig. 2.12

Let us consider a toroidal coil (Rowland ring) carrying a current I_0 in it. The ring is so constructed that its core can be easily replaced. If the current in the winding is constant the value of B will be greater when the core is in place than without it. The large value of B is due to magnetisation of the material of the core. If I_0 is the current through toroid of radius r_0, N_0 are the number of turns then, according to Ampere's law, the value of $\vec{B}$ without iron core:

$$\oint \vec{B}.d\vec{l} = \mu_0 I \quad ...(10)$$

$$(B)\ 2\pi r_o = \mu_o N_o I_o \quad ...(11)$$

Let us now put the core. The value of $\vec{B}$ determined experimentally is different when an iron core is placed than without it. As I_0 is the same then according to Ampere's law B must be same, which is not in agreement with experiment. *Thus Ampere's law fails.*

However, the value of $\vec{B}$ can be increased and equal to its value with core, even i the absence of the core by increasing the current in the toroid say by an amount I_{mo}. *The magnetisation of the iron core is thus equivalent in its effect on B to such a hypothetical current increase.* We modify the Ampere's law by adding magnetic current term I_m for fictitious amperian (or atomic) currents. To make it clear we can say as the Gauss's law was modified in the presence of dielectric, similarly the Ampere's law is modified in the presence or magnetic material, *i.e.*,

$$\oint \vec{B}.d\vec{l} = \mu_o (I + I_m) \quad ...(12)$$

or $$(B)\ 2\pi r_o = \mu_o N_o I_o + \mu_o (N_o I_{mo}) \quad ...(13)$$

In fact I_{mo} is related to the magnetisation $\vec{M}$, see eq. (10). *It is the value of the current which has the magnetic dipole moment equal to the magnetic moment produced due to the alignment of atomic dipoles.* If A is the area of the toroid, then the magnetic moment for the element dl of the toroid,

$$d\vec{m} = \vec{M}\ (Adl)$$

But magnetic moment is defined by,

$$m = NIA$$

where N are the number of turns, I is the current through the loop. For element dl, the number of turns

$$= \text{Number of turns per unit length Xdl}$$

$$= \frac{N_o}{2\pi r_o} dl$$

therefore the magnitude of magnetic moment, for element dl

$$M(Adl) = \frac{N_o}{2\pi r_o} dl.\ I_{mo} A$$

or $$N_o I_{mo} = M\ 2\pi r_o \quad ...(14)$$

Substituting this value of No Imo in equation (14), we get,

$$(B)\ 2\pi r_o = \mu_o (N_o I_o) + \mu_o (M\ 2\pi r_o)$$

or in general $\oint \vec{B}.d\vec{l} = \mu_o I + \mu_o \oint \vec{B}.d\vec{l}$

or, $$\oint \left(\frac{\vec{B} - \mu_o \vec{M}}{\mu_o} \right).d\vec{l} = I \qquad ...(15)$$

The quantity $\frac{\vec{B} - \mu_0 \vec{M}}{\mu_0}$ is given a special name *magnetic field strength* $\vec{H}$, hence,

$$\vec{H} = \frac{\vec{B} - \mu_o \vec{M}}{\mu_o} \qquad ...(16)$$

or, $$\vec{B} = \mu_o \vec{H} + \mu_o \vec{M} \qquad (17)$$

or $$\vec{B} = \mu_o \left(1 + \frac{\vec{M}}{\vec{H}} \right) \vec{H} \qquad ...(18)$$

Restatement of the Ampere's law:

Ampere's law, in the presence of magnetic material is given by eqns. (17) and (18), *i.e.,*

$$\oint \vec{H}.d\vec{l} = I \qquad ...(19)$$

where I is true current and does not include the magnetising current. The equation states the most important property of field intensity $\vec{H}$ – *the line integral of* $\vec{H}$ *around any closed path equals the total conduction current encircled regardless of the presence of magnetised material.*

To find $\vec{H}$ in our toroid case, let us apply Ampere's law as given in equation (20).

$$H\ (2\pi r_o) = N_o I_1$$

or, $$H = \left(\frac{N_o}{2\pi r_o} \right) I_o = n.I_o$$

where n are the number of turns per unit length.

Magnetising force or intensity $\vec{H}$

The relation $\oint \vec{B}.d\vec{l} = \mu I$ shows that the value of $\vec{B}$ depends on the permeability of the medium. From relation $\oint \vec{H}.d\vec{l} = I$, it can be seen that $\vec{H}$ does not depend on permeability m but only the currents. $\vec{H}$ is independent, therefore, of the medium in which the conductors are situated. It is this reason that $\vec{H}$ is regarded as being due to the currents only. $\vec{H}$ is then a "cause" which gives rise to a flux density $\vec{B}$ given by $\mu\vec{H}$, and so $\vec{B}$ is dependent on the medium used.

Because of this interpretation, $\vec{H}$ is often called the "magnetising force" or magnetising intensity. $\vec{H}$ is a vector in the direction of $\vec{B}$. The term magnetic intensity is not particularly appropriate since it implies that $\vec{H}$ is analogous to the electric field intensity $\vec{E}$, which is not precisely correct since in electric field $\vec{E}$ is defined in terms of force, whereas in magnetic field it is $\vec{B}$ that is defined in terms of force. Therefore, the name magnetising force is sometimes used for $\vec{H}$ which is more appropriate.

Unit of $\vec{H}$

The unit of $\vec{H}$ is obtained by equation (2.24) and is Ampere-turn per metre.

IMPORTANT CONCLUSIONS ABOUT $\vec{B}.\vec{H}$ AND $\vec{M}$

Here, we are summarising the previous results.

(1) First property of $\vec{B}$ is derived from Gauss's law, *i.e.*, div $\vec{B}$ = 0, which means $\vec{B}$ lines are closed circles and do not have sources or sinks. In other words absence of monopole.

(2) Second property of $\vec{B}$ is expressed by the Ampere's law $\oint \vec{B}.d\vec{l}$ = $m_0 I$, which means $\vec{B}$ is *non-conservative* field *i.e.*, do not have sources or sinks in contrast to the electric field $\vec{E}$ which originates from charges..

(3) We have restated Ampere's law as,

$$\oint \vec{H}.d\vec{l} = I \quad ...(1)$$

where I is the true conduction current. According to this relation $\vec{H}$ *is also a non-conservative field.*

By Stoke's theorem,

$$\oint \vec{H}.d\vec{l} = \int_S \text{Curl } \vec{H}.d\vec{S} \qquad ...(2)$$

And current,

$$I = \int_S \vec{J}.d\vec{S} \text{ (see eqn. 8.17)} \qquad ...(3)$$

By comparison of equations A, B and C we get,

$$\text{Curl } \vec{H} = \vec{J} \qquad ...(4)$$

This is the basic equation valid everywhere.

(4) We have established the relation

$$\oint \vec{M}.d\vec{l} = I_m$$

where I_m is amperian current, equation states that the line integral of the magnetisation around a closed path is equal to the total amperian currents through the surface encircled by the path.

Now using Stoke's theorem,

$$\oint \vec{M}.d\vec{l} = \int_S \nabla \times \vec{M}.d\vec{S}$$

and $$I_m = \int_S \vec{J}_m.d\vec{S}$$

therefore, $\nabla \times \vec{M} = \vec{J}_m$ ...(5)

which is the same as proved previously in equation (5)

(5) **Important point** to be noted is that sometimes the relation, $\vec{B} = \text{mo } (\vec{H} + \vec{M})$ is taken as,

$$\vec{B} = \vec{H} + \mu_0 \vec{M}$$

Then in RMKS, $\vec{H}$ will have the same unit as $\vec{B}$ *i.e.*, weber/metre^2. However in our relation $\vec{B} = \mu_0 (\vec{H} + \vec{M})$, $\vec{H}$ is expressed in the same unit as $\vec{M}$ *i.e.*, ampere per metre. *This unit is more convenient in engineering for designing transformers, magnets etc.*

Further, in our relation $\vec{B}$ and $\vec{H}$ are expressed in different units. There is no difficulty with this definition of $\vec{H}$ in RMKS units. Hence we prefer relation $\vec{B} = \mu_0 (\vec{H} + \vec{M})$ then to $\vec{B} = \vec{H} + \mu_0 \vec{M}$.

PERMEABILITY AND SUSCEPTIBILITY RELATIVE PERMEABILITY

We can measure $\vec{B}$, $\vec{H}$ and $\vec{M}$ for different magnetic materials For non-ferromagnetic that is, for paramagnetic and diamagnet materials it is found experimentally that $\vec{B}$ is directly proportional to $\vec{H}$

or $\quad \vec{B} \propto \vec{H}$

or $\quad \vec{B} = \mu_r \mu_o \vec{H}$ ...(1)

where μ_r (sometimes written as K_m) is the relative permeability of the magnetic medium. $\vec{B}$ is vector in the direction of $\vec{H}$. Putting the value of $\vec{B}$ from eqn. (1) in eqn. (2), we get,

$$\mu_r \mu o \vec{H} = \mu_o \vec{H} + \mu_o \vec{M}$$

or $\quad \vec{M} = (\mu_r - 1) \vec{H}$ ...(2)

Now in vacuum, the magnetisation $\vec{M}$ is zero, because there are no magnetic dipoles present, hence putting $\vec{M} = 0$, equation (2) reduces,

$$\vec{B} = \mu_o \vec{H} \qquad ...(3)$$

Since in vacuum $\vec{M} = 0$, hence from equation (4)

$$(\mu_r - 1) = 0$$

or $\quad \mu_r = 1$ for vacuum

Magnetic susceptibility

For para and diamagnetic substances the magnetisation is proportional to the magnetising field H,

$$M \propto H$$

or $\quad \vec{M} = \chi_m \vec{H}$ ...(4)

where χ_m is known as the magnetic susceptibility.

Since, $\chi_m = \dfrac{M}{H}$. It may be defined as ratio of magnetisation to the magnetising field. Comparing equation (3) and (4), we get

$$\chi_m = \mu_r - 1$$

or $\quad \mu_r = 1 + \chi_m$ ...(5)

The magnetic susceptibility of the medium is the measure of the capability of the medium to take up magnetisation. It is analogous to dielectric susceptibility.

Permeability

Substituting equation (5) in eqn. (4), we get,

$$\vec{B} = (1 + \chi_m)\, \mu_o\, \vec{H}$$

or $$\vec{B} = \mu\, \vec{H} \qquad ...(6)$$

where $\mu = \mu_o\,(1 + \chi_m)$, is the permeability of the medium, the dimensionless quantity,

$$\frac{\mu}{\mu_o} = 1 + \chi_m = \mu_r \text{ (by eqn 5)} \qquad ...(7)$$

or $$\mu = \mu_o \left(1 + \frac{M}{H}\right), \sin \chi_m = \frac{M}{H} \qquad ...(8)$$

where μ_r is relative permeability of the medium and is defined as the ratio of the number of field lines per square metre taken at right angles in the medium to that in air or free space.

The permeability μ *measures the degree of penetration of magnetic field through the substance or in other words it measures the capacity of the substance to take magnetisation.* In isotropic medium $\vec{M}$ and $\vec{H}$ are in the same direction, hence m is scalar. However, in non-isotropic media, $\vec{M}$ and $\vec{H}$ are, in general not in the same direction and m is not a scalar but becomes a nine components quantity or tensor. Therefore relation (9) is general relation and (10) is concise expression only for isotropic media.

Units

From equation (9), the units of μ

$$= \frac{\text{Units of B}}{\text{Units of H}}$$

$$= \frac{\text{Webers' metre}^2}{\text{Ampere / metre}}$$

$$= \frac{\text{Webers}}{\text{Ampere.metre}},$$

$$\text{since henry} = \frac{\text{Webers}}{\text{Ampere}}'$$

Magnetic Properties of Matter

$$= \frac{\text{Henrys}}{\text{Metre}}$$

or from equation (2.33), the unit of μ is same as mo

$$= \frac{\text{Weber}}{\text{Ampere} \cdot \text{metre}} = \frac{\text{Henrys}}{\text{metre}}$$

Since μ_r is ratio it is dimensionless quantity. From eqn. (2.30) since $\vec{M}$ and $\vec{H}$ are expressed in the same unit, *i.e.*, ampere/metre, the ratio, χ_m is dimensionless quantity.

REPRESENTATION OF $\vec{B}$, $\vec{H}$ AND $\vec{M}$

Since magnetisation occurs inside the materials. According to equation (1),

$$\vec{B} = \mu_0 \vec{H} + \mu_0 \vec{M}$$

At the boundary $\vec{H}$ is reversed. Thus within the magnets $\vec{M}$ and $\vec{H}$ point in opposite direction, as shown in Fig. 2.13. How $\vec{H}$ is reversed? Let us examine it as follows.

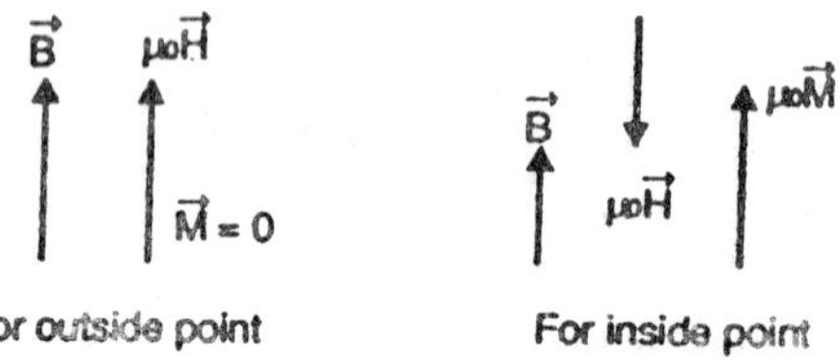

Fig. 2.13

Magnetic Scalar Potential

Taking the divergence of equation (1), *i.e.*,

$$\vec{B} = \mu_0 \vec{H} + \mu_0 \vec{M}$$

$$\nabla.\vec{B} = \nabla.\mu_0 \vec{H} + \nabla.\mu_0 \vec{M}$$

Since ∇, $\vec{B} = 0$, we have,

$$\nabla.\vec{H} = -\nabla.\vec{M} \qquad ...(2)$$

Since $\vec{M}$ and $\vec{H}$ having the dimensions of current/length, (ampere/metre). Then $\nabla.\vec{H}$ or $\nabla.\vec{M}$ have the dimensions as current per area (ampere/metre^2) which can be conveniently expressed as,

$$\frac{\text{Ampere.metres}}{\text{Metres}^2} = \frac{\text{Pole strength}}{\text{Volume}}$$

As we have proved that a current carrying coil is equivalent to a magnet (*i.e.*, IA = Q_{ml} where Q_m is the pole strength), hence dimensions of pole strength Q_m are (current × length), *i.e.*, (Amp. metre).

Therefore $\nabla\vec{H}$. or $\nabla.\vec{M}$ can also be expressed as pole strength per volume.

We can write,

$$\nabla.\vec{H} = -\nabla.\vec{M} = \rho_m \qquad ...(3)$$

where ρ_m = pole volume density $\left(\frac{\text{Ampere.metres}}{\text{metres}^2}\right)$ the quantity ρ_m is like the polarisation volume density ρ_p.

If the divergence of a vector field is not zero, means the field has a source, or place of origin. Let us recall the polarised dielectric case, *i.e.*, $\nabla.\vec{p} = \rho_p$, which indicates that the polarisation field originates on the polarisation charge at the dielectric surface. In an analogous manner equation (2) *indicates that the* $\vec{H}$ *field originates where the magnetisation field* $\vec{M}$ *ends and the* $\vec{H}$ *field ends where the* $\vec{M}$ *field originates.* This occurs at the surface of the substance. Thus, if we consider points entirely outside of the current regions, and choose a path which do not enclose any current, thus when no current is enclosed,

$$\oint \vec{H}.d\vec{l} = 0 \qquad \text{(by Ampere's law)}$$

Under this condition $\vec{H}$ *is conservative field* and can then be derived from a scalar magnetic potential function* (V_m).

Thus when path does not enclose current, the above relation by Stoke's theorem can be written as

$$\nabla \times \vec{H} = 0$$

Then $\vec{H}$ can be expressed as negative gradient of a scalar in the same manner as in electrostatics $\nabla \times \vec{E} = 0$, then $\vec{E} = -\nabla V$.

$$\vec{H} = -\nabla V_m \qquad ...(4)$$

Taking div. of (4), we get,

$$\nabla . \vec{H} = -\nabla^2 V_m$$

Therefore, from equation (5),

$$\nabla^2 V_m = \nabla . \vec{M} \qquad ...(5)$$

This indicates that magnetic scalar potential V_m is related to source of magnetisation. *This is the "Poisson's equation" solution of which gives the magnetic scalar potential in current free regions.*

Vector Potential

Taking the curl of the relation,

$$\vec{B} = \mu_o \vec{H} + \mu_o \vec{M}$$

$$\nabla \times \vec{B} = \mu_o \nabla \times \vec{H} + \mu_o \nabla \times \vec{M} \qquad ...(6)$$

when there is no magnetisation, equation (6) reduces to,

$$\nabla \times \vec{B} = \mu_o \nabla \times \vec{H}, \text{ [since } \nabla \times \vec{H} = \vec{J} \text{ by eqn.} \qquad ...(7)]$$

$$= \mu_o \vec{J} \qquad ...(8)$$

From relation (6), *the curl of* $\vec{M}$ *has the dimensions of current density (amp./metre²) and represents the equivalent current of density* $\vec{J}_m$ (amp./metre²) *flowing, in a very thin layer around the surface of a uniformly magnetised substance (as rod).*

Hence $\nabla \times \vec{M} = \vec{J}_m$ again the same conclusion as given in equation (7) and (8).

Thus equation (7) reduces,

$$\nabla \times \vec{B} = \mu_o (\vec{J} + \vec{J}_m) \qquad ...(9)$$

where $\vec{J}$ = actual current density, as in a current carrying wire (amp./metre²).

$\vec{J}_m$ = equivalent current density, as at the surface of a magnetised substance (bar) amp./metre²

Potential due to a Magnetisation

We have defined the vector potential as,

$$\vec{B} = \nabla \times \vec{A}$$

and, $\vec{A} = \frac{\mu_o}{4\pi}\int \frac{\vec{J}}{r} dv'$ see eqn. (10)

If both conduction and magnetisation currents are present, we have

$$A = \frac{\mu_o}{4\pi}\int_o \frac{\vec{J} + \vec{J}_m}{r} dv' \qquad \text{...(11)}$$

where, $\vec{J} = \nabla \times \vec{H}$ amp./metre^2

$\vec{J}_m = \nabla \times \vec{M}$ amp./metre^2

COMMENT ON THE FIELD $\vec{H}$

We have restated Ampere's law as,

$$\oint \vec{H}.d\vec{l} = I$$

From the discussion of the preceding section it is clear that $\vec{H}$ can be regarded as sum of two components.

(1) H_1 due to true currents and it is non-conservative and

$$\oint \vec{H}_1.d\vec{l} = I$$

By non-conservative means it does not have sources of sinks.

(2) However, if we consider the magnetisation and if H_c is the field due to magnetisation, then it is a conservative field that satisfies the relation,

$$\oint \vec{H}_c.d\vec{l} = 0$$

In fact H_c represents the conservative field due to fictitious point charges those are assumed when substance is magnetised and then we have the magnetic scalar potential only in current free regions (As when $\oint \vec{H}.d\vec{l} = 0$, we express $\vec{E} = -\nabla V$).

BOUNDARY CONDITIONS AT THE SURFACE OF A MAGNETIC MATERIAL

Let us determine the relation between the values of $\vec{B}$ and $\vec{H}$ inside a magnetic material to the values of $\vec{B}$ and $\vec{H}$ just outside the material. The relation is obtained by two basic laws, the first is that the field $\vec{B}$ consists of closed lines, *i.e.*, the flux of $\vec{B}$ through a closed surface is zero which is precisely Gauss's law,

$$\nabla.\vec{B} = 0$$

or $\oint \vec{B}.d\vec{l} = 0$ ...(1)

The second basic law is Ampere's law, *i.e.*,

$$\oint \vec{H}.d\vec{l} = I \quad ...(2)$$

Applying equation (1) to the cylindrical surface of height h shown in Fig. 2.14 we obtain,

$$B_{n1} \nabla S - B_{n2} \nabla S = 0 \quad ...(3)$$

where Bn1 is the average component of $\vec{B}$ normal to top face of the cylinder and B_n^2 is the average component of $\vec{B}$ normal to the lower face in inward direction, that is B_n^2 and ∇S are oppositely directed.

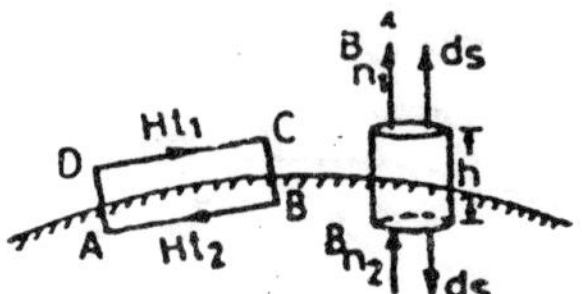

Fig. 2.14

The integral of $\vec{B}$ over the curved surface of the cylinder is neglected because by making height h approaching zero, integral zero even though there may be finite of $\vec{B}$ normal to the surface, by eq. (4)

$$B_n^1 = B_n^2 \quad ...(4)$$

Therefore, the normal component of $\vec{B}$ inside the substance is the same as the normal component of $\vec{B}$ just outside the substance. If the boundary separates two media of different permeability, even then equation (4) is valid. *Thus, the normal component of the* $\vec{B}$ *is continuous across the boundary between the two media.*

Let us now apply Ampere's law to the path ABCD, AB = – ∇_x. Let the average value of $\vec{H}$ tangent to the boundary for upper half is H_{t1} and for lower half path H_{t2}, then according to Ampere's law,

$$H_{t1} \nabla_x - H_{t2} \nabla_x = I$$

where I is current enclosed by ABCD. The integral of $\vec{H}$ over the segments BC and DA becomes zero even though a finite field may exist normal to the boundary by choosing BC and DA as small as possible and approach thing zero. The second term is negative because H_{t2} and ∇_x are oppositely directed.

$$H_{t1} - H_{t2} = \frac{I}{\nabla x} = I' \text{ amp/metre} \qquad ...(5)$$

where I' is the linear density of current *i.e.*, current per unit length flowing in an infinitesimally thin sheet at the surface. According to equation (5), the change in tangential component of $\vec{H}$ across a boundary is equal in magnitude to the linear current density. If I' = 0

$$H_{t1} = H_{t2} \qquad ...(6)$$

According to equation (6) the tangential component of $\vec{H}$ inside the magnetic material is the same as the tangential component of $\vec{H}$ just outside the material. If the boundary is separating two media of different permeability, even then equation (6) is valid.

Thus, the tangential components of $\vec{H}$ are continuous across the boundary between two media provided the boundary has no current sheet of infinitesimal thickness.

CLASSIFICATION OF MAGNETIC MATERIALS

Faraday divided the magnetic materials into three groups in terms of susceptibility χ_m and relative permeability μ_r.

Paramagnetic Substances

For paramagnetic substances $\vec{M} = \chi_m \vec{H}$ and susceptibility cm is positive and less than unity and the relation $\mu_r = 1 + \chi_m$ indicates that relative permeability is slightly greater than unity. Susceptibility χ_m is independent of the field, when $\vec{H} = 0$ $\vec{M} = 0$ in this case. Paramagnetic substance when placed in magnetic field *acquire a feeble induced magnetism in the same direction as the external field as* $\chi_r > 1$. In a non-uniform field it will experience an attractive force toward the strongest point. The susceptibility cm varies inversely with the absolute temperature, mathematically,

$$\chi_m \propto \frac{1}{T}$$

or

$$\chi_m T = \text{Const.} \qquad ...(1)$$

Examples:

Some examples of paramagnetic substances are oxygen, platinum, aluminium, chromium, manganese and solutions of iron and nickel.

Diamagnetic Substances

Such substances acquire feeble induced magnetism in a direction opposite to the external field. In non-uniform field, they have tendency to move from stronger to weaker parts of the field. The magnetisation M is proportional to the magnetising field,

$$\vec{M} = \chi_m \vec{H}$$

Here, the susceptibility χ_m is independent of the temperature. χ_m is negative and relative permeability μ_r is slightly less than unity *i.e.,* $\mu_r < 1$. *Examples* of diamagnetic substances are bismuth, antimony, copper, mercury, gold, water, air, hydrogen and alcohol.

Ferromagnetic Substances

For ferromagnetic substances, the susceptibility χ_m is positive and is larger than para magnetic substances and the relative permeability is also much greater than unity. However, the relation $\vec{M} = \mu_c \vec{H}$ (*i.e.,* the magnetisation proportional to the field) does not hold for ferromagnetic materials. The ferromagnetic substances having spontaneous magnetisation even without the field, when $\vec{H} = 0$, $M \neq 0$ in this case. Due to greater value of μ_r, they are able to acquire high degree of magnetisation even they can develop north and south pole.

Distinction between Para and Ferro

Susceptibility decreases with increase in temperature. If the temperature is increased above a certain critical value, called Curie temperature the substance loses its ferromagnetism, that is, the spontaneous magnetisation disappears and substance becomes paramagnetic. Above the Curie temperature the susceptibility follows the Curie Weiss law, *i.e.,*

$$m = \frac{C}{T - \theta} \quad ...(2)$$

where C is Curie-constant and temperature θ is Curie-temperature. *Hence we conclude that all ferromagnetism are basically paramagnetic but all para are not ferromagnetics. Above the curie temperature substance is paramagnetic and below it, it is ferromagnetic.*

Examples:

The five elements iron, cobalt, nickel, Gadolinium and dysprosium and variety of alloys of these and other elements.

MAGNETISATION CURVES (B – H CURVES)

The permeability μ of a substance is given by equation (2) and (3), *i.e.,*

$$\mu = \frac{B}{H} = \mu_r \mu_o$$

But the permeability μ of the ratio B/H is not a constant for ferromagnetic materials in other words B is not proportional to $\vec{H}$. Therefore, in order to know the relation of $\vec{B}$ to $\vec{H}$ a graph showing the variation of B with H is plotted by taking B as ordinate and H as abscissa. The line or curve showing B as a function of H as such B-H curve is called magnetisation curve.

Ballistic Method for Plotting H-H Curve

To plot a magnetisation curve for an iron sample, take the sample in the form of a ring. A primary coil P is wound uniformly and closely spaced over the ring, forming an iron-cored toroid as in Fig. 2.15.

The ends of the primary coil are connected to a battery B_1, through a resistance box R_1 and a reversing key K_1.

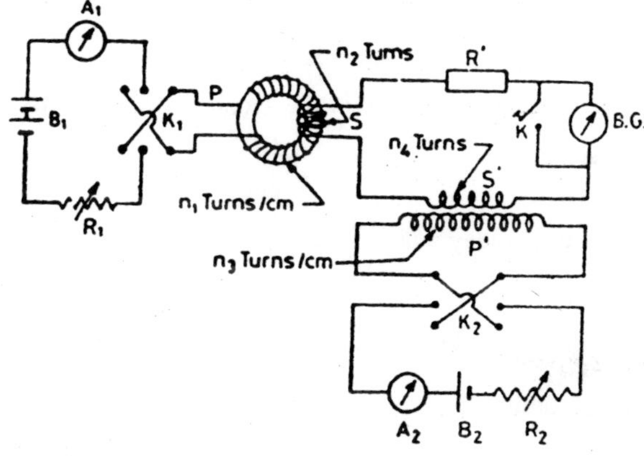

Fig. 2.15

The small coil S called the secondary coil, is wound over the primary coil P and is connected to ballistic galvanometer through a rheostat R' and secondary coil S'. The secondary coil S' itself wound over a primary P', which is further connected to battery B_2 through a reversing key K_2 (a commutator) for calibrating the B.G.

First the reversing key K_1 is switched on. A Transformers and A.C. Bridges I passes through the primary coil P. If the number of turns per unit length are n_1. Then, the value of H applied to the ring is,

$$H = n_2 I \text{ amp/metre or amp. turns/metre} \quad ...(3)$$

(Since, for toroid $B = \mu_o n_1 I$ and there is no magnetisation so far hence $\vec{M} = 0$, then relation $\vec{B} = \mu_o \vec{H} + \mu_o \vec{M}$ is $B = \mu_o H$, hence $\vec{H} = \frac{B}{\mu_o} = n_1 I$).

This value of H applied to the ring may be called the "magnetising force". The magnetic induction B in the ring may be regarded as the result of applied field H and is measured by placing the secondary coil S over the ring. The magnetisation of the specimen develops a magnetic flux density (or magnetic induction) B inside the ring which is transferred to the secondary S. This transfer of flux to the secondary coil causes an induced emf.

Suppose the flux density developed in the ring in the time dt = dB

Therefore, the flux developed pet turn = AdB where A is the area of cross-section of the ring.

The flux linked with secondary = $dBAn_2$ where n_2 are the number of turns of the secondary.

Since initially flux in the secondary is zero.

The change in magnetic flux in the secondary = $d\phi = dB.An_2$

Rate of change of flux $\frac{d\phi}{dt} = An_2 \frac{dB}{dt}$

According to Faraday's law $\frac{d\phi}{dt}$ is equal to the induced emf in the secondary.

$$\xi = An_2 \frac{dB}{dt} \text{ volts}$$

And corresponding current in the secondary,

$$I_2 = \frac{\xi}{R} = \frac{An_2}{R}\frac{dB}{dt} \text{ amps}$$

where R is the total resistance of the galvanometer circuit.

Since current is rate of flow of change, hence,

$$\frac{dq}{dt} = \frac{An_2}{R}\frac{dB}{dt}$$

$$dq = \frac{An_2}{R} dB \text{ Couls}$$

Since initially magnetic induction B was zero it rises as current increases from zero to a saturation value B. Thus total charge developed in the secondary,

$$q = \int_O^B dq = \int_O^B \frac{An_2}{r} dB = \frac{An_2R}{R} \text{ Coul}$$

Now if θ is the first throw in B.G. the,

$$q = K\theta\left(1+\frac{\lambda}{2}\right), \qquad \text{K is B.G. constant,}$$

from these two expressions, we get'

$$\frac{An_2B}{R} = K\theta\left(1+\frac{\lambda}{2}\right) \qquad ...(4)$$

Determination of Galvanometer Constant

The value of K can be determined by the use of a standard solenoid. A known current I' is passed through the primary of the standard solenoid and suddenly this will cause change in magnetic flux is calculated as follows. The magnetic field in the primary P' = n_2I' where n_3 are the number of turns per unit length of the primary P'.

This creates a magnetic flux through the secondary,

$$\phi' = \mu_o n_3 I' \propto n_4$$

where ∝ is the area of cross-section of the secondary S' and n_4 are the number of turns in the secondary S'. This change in flux will give rise to an induced emf in the secondary hence proceeding in the manner as above, we get,

$$\frac{\mu_o n_3 I' \propto n_4}{R} = K\theta'\left(1+\frac{\lambda}{2}\right) \qquad ...(5)$$

where q' is the deflection produced. The constant K can be eliminated from eqns. (4) and (5) *i.e.,* dividing eqn. (4) by (5),

$$\frac{Bn_2A}{\mu_o n_3 I' \propto n_4} = \frac{\theta}{\theta'}$$

or

$$B = \frac{\mu_o n_3 I' \propto n_4}{An_2}\frac{\theta}{\theta} \text{ weber/metre}^2 \qquad ...(6)$$

and H = n_1I amp/metre

These eqns. give the value of B and H in terms of known quantities. *Experimentally the B-H curve may be plotted either (1) by steps or (2) by reversal.*

(1) *Step method*: In this method the current is increased in steps and ballistic throw is observed. The total current and magnetic field are obtained by adding together the individual values in steps.

(2) *Reversal method*: This method is better because when the current is reversed from + I to – I, the value of B is double as the flux change will be double, hence the ballistic throw q is also doubled. Further, when calibrating B-G, the reversal of current I' doubles the flux density B, consequently the ballistic throw q' is also doubled. Therefore, the value of B does not change as the values of equations (1) and (2) are doubled, hence the ratio of equations (1) and (2) given in equation (6) for B remains unchanged.

DETERMINATION OF SUSCEPTIBILITY C_M AND PERMEABILITY μ

From equation (1),

$$\vec{B} = \mu\vec{H} \quad ...(1)$$

where $\mu = \mu_o (1 + \chi_m)$

$$\therefore \quad \vec{B} = \mu_o (1 + \chi_m)\vec{H} \quad ...(2)$$

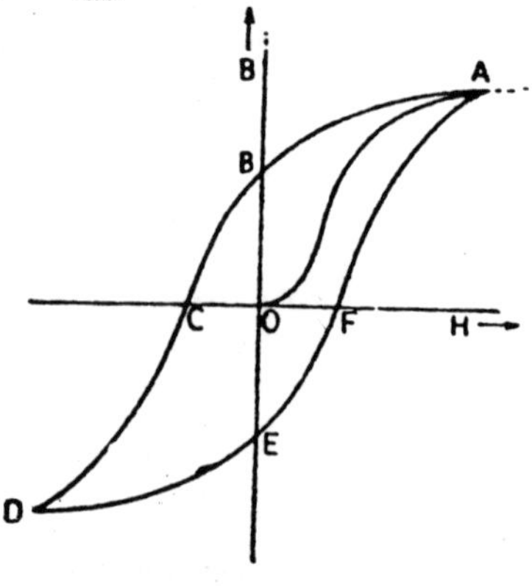

B
A
B
C
O
F
H
E
D

Fig. 2.16

Thus, by knowing $\vec{B}$ and $\vec{H}$, we can calculate the susceptibility cm by using equation (2) and B from equation (1). The $\vec{B}$-$\vec{H}$ curve is shown in Fig. 2.16.

$\vec{B}$-$\vec{H}$ Curve

Since B and H are known, we can also find the value of magnetisation M by the relation,

$$\vec{B} = \mu_o (\vec{H} + \vec{M})$$

Then a curve between $\vec{M}$ and $\vec{H}$ can be drawn it is called $\vec{M}$-$\vec{H}$ curve and is identical to $\vec{B}$-$\vec{H}$ curve.

Variation of Permeability

Although $\mu = \frac{B}{H}$, for ferromagnetic materials, μ depends on the magnetising field H. *Various kinds of permeability can be defined from a hysteresis loop.*

1. **Ordinary or normal permeability**. It is the ratio B/H for a point on the normal magnetisation curve, *i.e.*, ratio B/H for the tip of the loop.
2. **Differential permeability**. It is the slope $\frac{dB}{dH}$ of the magnetisation curve.
3. **Initial permeability**. The value of $\frac{dB}{dH}$ for H = B = 0 is called the initial permeability.

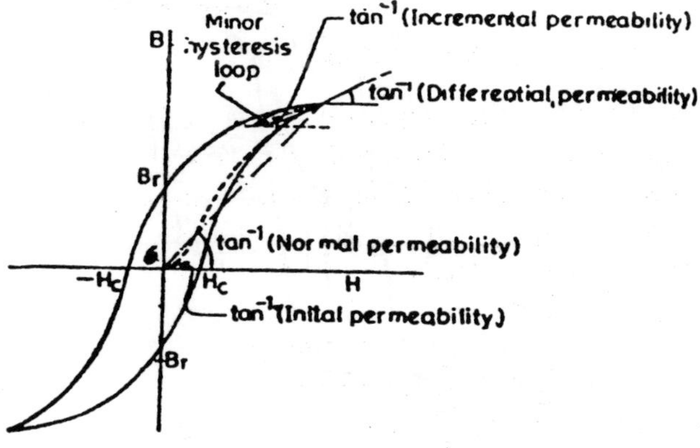

Fig. 2.17

4. **Incremental permeability**. If a steady value of H is applied to a specimen and then small alternating H is superimposed a minor hysteresis loop is obtained that has its centre near the original point on B-H graph. The slope of the line connecting tips of this minor loop is the incremental permeability. All these are shown in Fig. 2.17.

CYCLE OF MAGNETISATION HYSTERESIS

Let us consider an unmagnetised ferromagnetic substance an iron bar in a magnetising field. The magnetising field is slowly increased as in the case of B-H curve. The magnetisation $\vec{M}$ increases as represented by the part OA, as shown in Fig. 2.18. After A any further increase in H would not increase $\vec{M}$. That is the substance reaches the "*magnetic saturation*" $\vec{M}$. In other words, all elementary dipoles have been aligned along the field direction. Now the magnetic field H is gradually decreased to zero. The magnetisation decreases but does not become zero when H = 0. The magnetisation $\vec{M}$ is represented by OB. This value of $\vec{M}$ corresponding to H = 0 is called "*Residual Magnetism*" M_r or "*Retentivity*" or "*Remanence*". The remanence gives the state of permanent magnetisation of the sample. If we consider the B-H curve, the value of B corresponding to H = 0 *i.e.*, after reducing the H from saturation to zero is known as "*remanence induction*" B_r.

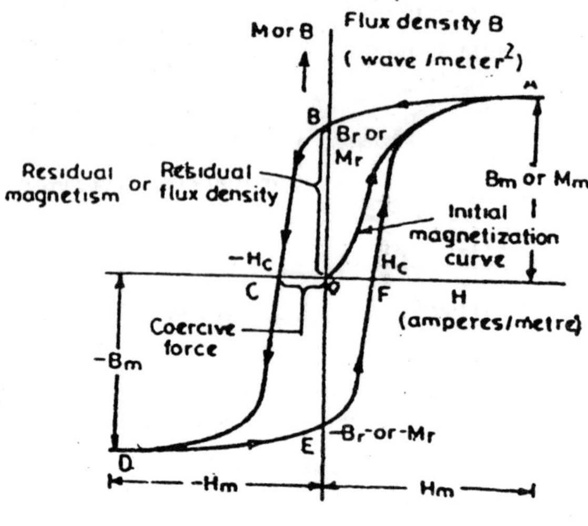

Fig. 2.18

If the direction of the field H is now reversed and its magnitude gradually increased, the curve BCD is obtained. The residual magnetism vanishes at C for the value OC of magnetic field in the negative direction. So *OC gives the magnetic field H needed to reduce the magnetisation to zero or in other words this is the magnetising force required to completely demagnetised the sample.*

This is called the "*coercive force*" or "*coercivity*" Hc. The three quantities referred above, saturation value, the residual magnetism together with the area of the hysteresis loop are able to explain up to some extent the nature of magnetic material.

Hysteresis

If we further increase H in the negative direction up to the saturation point D symmetrical to A. Now again, all the dipoles are oriented along the field direction but opposite to their direction of orientation at the saturation point A. If we now gradually reduce H to zero and reverse its direction at E, *i.e.*, restore its original direction. Then increasing the field again in steps, we reach back to the point A along the path DEFA. We trace out a path ABCDEFA is called the "hysteresis loop" and is characteristic of the magnetic properties of the material. It is clear from M-H curve shown in Fig. 2.12 that the descending part of the curve always lies above the ascending part. *Hence the magnetisation $\vec{M}$ or $\vec{B}$ (flux density) will become zero at a 'later point'* of the curve than the field H. It means the *magnetisation lags behind the magnetising field. This lag of magnetisation behind the magnetising field is called hysteresis.*

SOLVED EXAMPLES

Example 1:

If, H = 1000 amp./metre then calculate the magnetic potential difference between the ends of the bar and total flux.

Solution:

$V_m = \int \vec{H}.d\vec{l}$, as H is along the length of the bar, then $\vec{H}.d\vec{l}$ = Hdl.
Magnetic potential between the ends,

$$V_m = \int_0^{0.1} Hdl = 1{,}000 \times 0.1 = 1000 \text{ amp.}$$

The flux density $B = \mu_1 H = 500\mu_o \times 1000$ weber/metre^2

$= 6.28 \times 10^{-1}$ weber/metre^2

The total flux = BA

$= 6.28 \times 10^{-1} \times 15 \times 10^{-4}$

$= 9.4 \times 10^{-4}$ weber

Example 2:

In example 2.3 if B = 1 weber/metre^2, then calculate the potential difference between the ends of blocks.

Solution:

From boundary condition the normal component of B is continuous in block 1

$$H_1 = \frac{B}{\mu_1} = \frac{1}{500\mu_o} = 1.59 \times 10^3 \text{ amp.metre}$$

and potential difference

$$V_{m1} = H_1 l_1 = 1.59 \times 10^3 \times 0.1 = 159 \text{ amp.}$$

In block 2,

$$H_2 = \frac{B}{\mu_2} = \frac{1}{20000\mu_o} = 3.97 \times 10^2 \text{ amp/metre}$$

and potential difference V_m between the end faces of block 2,

$$V_{m2} = H_2 l_2 = 3.7 \times 10^2 \times 0.2 = 79.4 \text{ amp.}$$

The total potential difference across both blocks is then given by,

$$V_m = V_{m1} + V_{m2}$$

$$= 159 + 79.4 = 238.4 \text{ amp.}$$

Example 3:

Find the total reluctance and permeance between the ends of parallel-connected rectangular iron $\mu_l = 500\ \mu_o$ and $\mu_2 = 2000\ \mu_o$

Solution:

Since the blocks are connected in parallel, it is convenient to calculate the total permeance first.

$$\text{Permeance of block 1} = \frac{\mu_1 A_1}{l}$$

$$= \frac{500 \times 4\pi \times 10^{-7} \times 20 \times 10^{-4}}{0.2}$$

$= 6.28 \times 10^{-6}$ henry

$$\text{Permeance of block 2} = \frac{\mu_2 A_2}{l}$$

$$= \frac{2000 \times 4\pi \times 10^{-7} 10 \times 10^{-4}}{0.2}$$

$= 12.6 \times 10^{-6}$ henry

Total permeance = $(6.28 + 12.6)\ 10^{-6}$ henry

$= 1.89 \times 10^{-5}$ henry

The total reluctance is given by

$$= \frac{1}{1.89 \times 10^{-5}} = 5.3 \times 10^{4} \frac{1}{\text{henry}}$$

Example 4:

In the above example 2.6 if H = 1000 amp/metre, calculate the total flux.

Solution:

Since from boundary condition the tangential component of H is continuous that H is same in both blocks, the,

$$B_1 = \mu_1 H = 500\ \mu_o \times 1000 = 0.628 \text{ weber/metre}^2$$

$$\phi B_1 = B_1 A_1 = 0.628 \times 20 \times 10^{-4} = 1.26 \times 10^{-3} \text{ weber}$$

and $B_2 = \mu_2 H = 2000\ \mu_o \times 1000 = 2.52$ weber/metre^2

$$\phi B_2 = B_2 A_2 = 2.52 \times 10^{-3} \text{ weber}$$

Total flux $= \phi B_1 + \phi B_2 = 3.78 \times 10^{-3}$ weber

Example 5:

An iron ring have a cross-sectional area 10 cm^2 and air gap of width d = 2 mm and mean length $l = 2\pi R$ = 60 cm including the air gap. Find the number of turns required to produce flux density, B = 1 weber/metre^2, $\mu_{iron} = 10^{-3}$ henry/metre.

Solution:

From equation (2.65),

$$\text{Total flux} = \frac{\text{mmt}}{\text{Re luc tan ce}}$$

$$\phi_B = \frac{NI}{\dfrac{d}{\mu_o A} + \dfrac{l-d}{\mu A}}$$

Flux density, $\phi_B = BA = \dfrac{NI}{\dfrac{d}{\mu_o A} + \dfrac{l-d}{\mu A}}$

or $\quad NI = BA\left(\dfrac{d}{\mu_o A} + \dfrac{l-d}{\mu A}\right)$

$$= B\left(\frac{d}{\mu_o A} + \frac{l-d}{\mu}\right)$$

$$= 1 \times \left(\frac{2\times 10^{-6}}{4\pi\times 10^{-7}} + \frac{598\times 10^{-6}}{10^{-3}}\right)$$

$= 2188$ amp. turns.

Example 6:

An iron rod of length 50 cm and cross-sectional area 4 sq. cm. is in the form a closed circular ring. If the permeability of iron is 65×10^{-4} h/m. Calculate the number of ampere turns required to produce a flux of 4×10^{-4} weber.

Solution:

$$\text{Reluctance} = \frac{l}{\mu A} = \frac{0.5}{65\times 10^{-4} \times 4\times 10^{-4}}$$

$= 1.92 \times 10^5$ amp. turns/weber

mmf = NI

$$\text{Flux} \quad \phi_B = \frac{\text{mmf}}{\text{Reluctance}}$$

$$4 \times 10^{-4} = \frac{NI}{1.92\times 10^5}$$

NI = 77 amp turns

Example 7:

The magnetic susceptibility of medium is 948×10^{-11}. Calculate the permeability (or absolute permeability) and relative permeability.

Solution:

Hint: $\xi_m = 948 \times 10^{-11}$

Absolute permeability

$$\mu = \mu_o (1 + \xi_m)$$

$$= 4\pi \times 10^{-7} (1 + 948 \times 10^{-11}) \text{ henry/m}$$

$$= 4\pi \times 10^{-7} \times 949 \times 10^{-11} \text{ henry/m}$$

$$= 1.186 \times 10^{-14} \text{ henry/metre}$$

and relative permeability,

$$\mu_r = \frac{\mu}{\mu_o}$$

$$= \frac{4\pi \times 10^{-7} \times 949 \times 10^{-11}}{4\pi \times 10^{-7}}$$

$$= 949 \times 10^{-11}.$$

Example 8:

In the row land ring 2.0 amp current (i_o) is passing in the windings and the number of turns per unit length (n) in the toroid are 10 turns/cm. B measured is 1.00 weber/metre^2. Calculate (a), $\vec{H}$, (b) $\vec{M}$ and (c) the magnetising current, both when the core is present and when it is removed. (d) What is μ_r^2.

Solution:

(a) As $\vec{H}$ is independent of the core material and is given by Ampere's law

$$\oint \vec{H}.d\vec{l} = I$$

$\vec{H}$ is tangential to l

$$H \oint dl = NI$$

where N are the total number of turns on the toroidal windings and NI is enclosed current.

$$H_l = NI$$

$$H = \frac{N}{l} I = NI$$

$$= (10 \text{ turns/cm}) (2.0 \text{ amp})$$
$$= 2.00 \text{ amp} \times 10 \times 100 \text{ turns/metre}$$
$$= 2000 \text{ amp/metre}$$

(b) The magnitude of $\vec{M}$ is obtained by relation

$$\vec{B} = \mu_o \vec{H} + \mu_o \vec{M}$$

$$M = \frac{B - \mu_o H}{\mu_o}$$

$$= \frac{1 - 4\pi \times 10^{-7} \times 2000}{4\pi \times 10^{-7}}$$

(c) The effective magnetising current from Eq. (2.19a),

$$i_m = M.\left(\frac{2\pi r}{N}\right) = M/n$$

$$= \frac{7.9 \times 10^5 \text{ amp / metre}}{2.0 \times 10^3 \text{ turns / metre}}$$

$$= 390 \text{ amp.}$$

Thus, an additional current of this amount in the windings would produce the same value of B in the absence of a core as that obtained with the core in place.

(d) The permeability

$$\mu_r = \frac{B}{\mu_o H}$$

$$= \frac{1 \text{ weber / metre}^2}{(4\pi \times 10^{-7} \text{ web / amp. metre})(2 \times 10^2 \text{ a mp / met})}$$

$$= 397$$

EXERCISES

1. Define permeability and magnetic susceptibility. Show that

$$\mu = \mu_o (1 + \chi_m)$$

2. Define magnetic scalar and vector potentials. Show that Poisson's equation is

$$\nabla^2 V_m = \nabla . \vec{M}$$

and vector potential due to a magnetised material is

$$\vec{A} = \frac{\mu_o}{4\pi}\int_v \frac{\vec{j}+\vec{j}_m}{r}\,dv'$$

3. Distinguish dia, para-and ferro-magnetic materials. Give the Langvin theory of diamagnetism.
4. What are atomic currents? How will you distinguish between magnetic and non-magnetic materials?
5. Prove the magnetic due to orbital motion of an electron must be an integral multiple of eh/4πm.
6. What do you mean by magnetisation? What is difference between uniform and non-uniform magnetisation? Show that for non-uniform magnetisation

   $$\nabla \times \vec{M} = \vec{J}$$
7. Derive the relation between the resultant angular momentum and the resultant magnetic moment of atom. What is gyromagnetic ratio?
8. What are magnetisation curves? Explain residual magnetism, coercive force and hysteresis.
9. What is meant by hysteresis and cycle of magnetisation? Prove that
   (i) the area of $\vec{B}$-$\vec{M}$ cycle denotes the energy dissipated per cc of the magnetisation.
   (ii) or the energy dissipated is m_0 times the area of $\vec{M}$-$\vec{H}$ curve.
10. Describe the method (B.G.) of tracing of the hysteresis curve for a sample of iron in the form of an anchor ring.
11. Explain the magnetic circuit, magnetomotive force and reluctance. How a magnetic circuit differs from an electric circuit?
12. Distinguish between para. and ferromagnetic materials. Discuss the Langevin theory para-magnetism.
13. What is Curic Weises law?
14. Give domain theory of ferro-magnetism.
15. An iron ring of 10 cm radius and 5 sq. cm. cross-section has an air gap of 2 mm width. A coil is would uniformly over the ring.

A flux of 10^{-3} weber is produced in the gap. If a current of 1.82 amp. passes through the coil, find the total number of turns in the coil if the permeability of the core is 65×10^{-4} henry/metre.

16. An iron ring of radius 10 cm and area of cross-section 5 sq. cm. has an air gap 1 cm wide. The ring is would uniformly with coil of 900 turns. A flux of 10^{-4} weber is required in the air gap. Find the strength of current required to produce. Given miron = 65 $\times$ 10^{-4} henry/metre.

17. Find the number of ampere turns needed to produce a flux of 40,000 Maxwells through a closed soft-iron ring formed of a rod of square cross-section of side 2 cm, bent into a circle of outside diameter of 22 cm. Find also the number of amp. turns required for the same flux if the ring be cut into two halves each being separated by an air gap of 1 mm. Given that the permeability of iron is 1900. Assume that there is no leakage of flux (108 Maxwell = 1 weber).

18. Give a short account of our present idea about dia-para-and ferromagnetism.

19. How the Ampere's law is modified in presence of magnetic materials?

    Establish the relation $\oint \vec{H}.d\vec{l} = I$.

3

The Magnetic Field Due to Steady Currents

INTRODUCTION

We have introduced the concept of magnetic flux density and represented the magnetic field by a vector $\vec{B}$. Then calculated the forces and torques on current carrying wires. We know the cause of $\vec{B}$ that the current flowing through wire. How the $\vec{B}$ is related to current and the distance from the wire? We would like to know how various factors affect the magnitude and direction of $\vec{B}$. To be more specific, the objective is to formulate the equation from which $\vec{B}$ may be calculated if the current configuration is know.

THE SOLID ANGLE FORM OF BIOT-SAVART LAW AND THE MAGNETIC SCALAR POTENTIAL

The Biot-Savart law can be written as in the case of closed current loop,

$$\vec{B} = \frac{\mu_0 I}{4\pi} \oint \frac{d\vec{l} \times \hat{r}}{r^2}.$$

We now wish to show that the magnetic field near a current loop is proportional to the gradient of the solid angle subtended by the loop at the point where $\vec{B}$ is to be found. Let us displace the point P in Fig. 3.1 by an infinitesimal distance $d\vec{r}$ and find the value of $\vec{B}.\, d\vec{r}$

$$\vec{B}.\, d\vec{r} = \frac{\mu_0 I}{4\pi}\left(\oint \frac{d\vec{l} \times \hat{r}}{r^2}\right).\, d\vec{r}$$

The value of R.H.S. remains the same whether we first evaluate the integral then take the dot product with $d\vec{r}$ or we first take the product of $d\vec{l} \times \hat{r}/r^2$ with $d\vec{r}$ and then evaluate the integral around the whole circuit. Hence we can write,

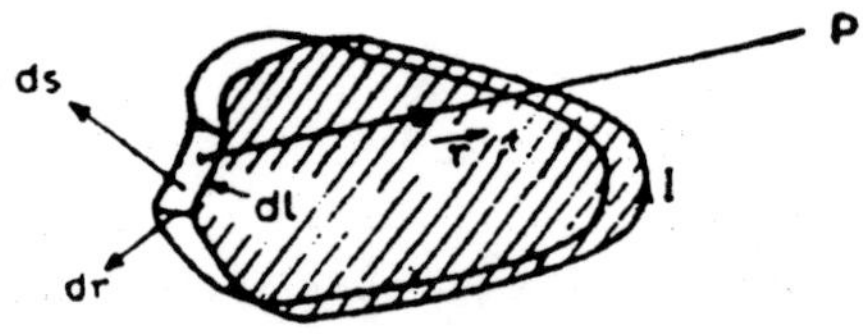

Fig. 3.1

$$\vec{B}.\,d\vec{r} = \frac{\mu_0 I}{4\pi} \oint \frac{(d\vec{l} \times \hat{r}).d\vec{r}}{r^2}$$

By the rules of triple scalar product *i.e.*,

$$(\vec{A} \times \vec{B}).\,\vec{C} = (\vec{B} \times \vec{C}).\,\vec{A} = (\vec{C} \times \vec{A}).\vec{B}$$

$$(d\vec{l} \times \hat{r}).\,d\vec{r} = (d\vec{r} \times d\vec{l}).\,\hat{r}$$

Therefore,

$$\vec{B}.\,d\vec{r} = \frac{\mu_0 I}{4\pi} \oint \frac{(d\vec{r} \times d\vec{l}).\hat{r}}{r^2}$$

$$= \frac{\mu_0 I}{4\pi} \oint \frac{(d\vec{l} \times (-d\vec{r})).\hat{r}}{r^2}$$

$$(\because \vec{A} \times \vec{B} = -\,\vec{B} \times \vec{A})$$

Now. $d\vec{l} \times (-d\vec{r}).\,\hat{r}/r^2$ *can be interpreted in terms of solid angles. The interpretation is clear, when the point P is displaced by an amount dr, the solid angle* Ω *subtended by the circuit at P changes by* $d\Omega$. *If we keep P fixed and move the entire current or in other words the each point of the circuit is displaced by an amount – dr, even then the change in solid angle is* $d\Omega$.

The quantity $d\vec{l} \times (-d\vec{r})$ is a vector representing the small area $d\vec{S}$ shown as a ribbon surface in Fig. 3.2 *i.e.*, $d\vec{S} = d\vec{l} \times (-d\vec{r})$ The projection of this area on a plane normal to r is given by

$$d\vec{l} \times (-d\vec{r}).\,\hat{r}$$

Therefore, $$\frac{d\vec{l} \times (-d\vec{r}).\,\hat{r}}{r^2} = \frac{d\vec{S}.\hat{r}}{r^2} = d\Omega$$

where $d\Omega$ represents the element of solid angle subtended by the area dS at P. Thus, the integral over the closed loop gives the entire solid angle subtended by ribbon. This solid angle is shown in Fig. 3.2. Here, it represents the decrease of the entire solid angle Ω subtended at P, that occurs when P moves a distance $d\vec{r}$ away from the loop moves $-d\vec{r}$ away from P Therefore, we write,

$$\oint \frac{d\vec{l} \times (-d\vec{r}).\ \hat{r}}{r^2} = -\ d\Omega.$$

Since Ω is a scalar function,

$$d\Omega = \vec{\nabla}\ \Omega . d\vec{r}$$

therefore, $$\oint \frac{d\vec{l} \times (-d\vec{r}).\ \hat{r}}{r^2} = -\ \nabla\ \Omega . d\vec{r}$$

Hence, $$\vec{B}.\ d\vec{r} = -\frac{\mu_0 I}{4\pi} \nabla\ \Omega . d\vec{r}$$

or $$\vec{B} = -\frac{\mu_0 I}{4\pi}\ \nabla\ \Omega = -\ \nabla \left(\frac{\mu_0 I \Omega}{4\pi}\right) \quad ...(1)$$

This Eqn. is the solid angle form of Biot-Savart law or the Ampere's law. The direction of $\vec{B}$ is that of $-\nabla\ \Omega$, so that $\vec{B}$ points away from the loop along its positive normal Eqn. (1) is the solid angle form of Ampere law. Now as we know that Ω is a scalar function. It is clear from eqn. (1) that the magnetic induction $\vec{B}$ is expressed in terms of scalar function Ω (x, y, z) of the coordinate of P which is analogous to the relation of electrostatics,

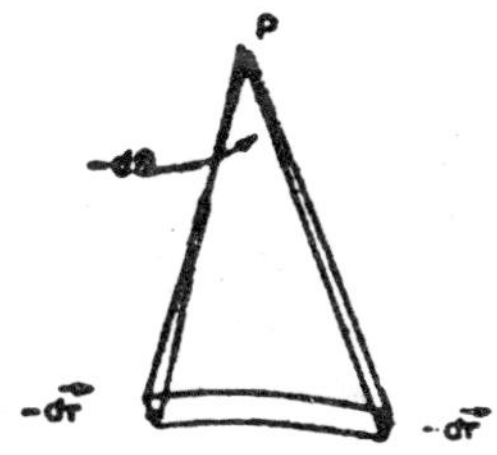

Fig. 3.2

$$\vec{E} = -\ \nabla\ V$$

We can write Eqn. (1) as

$$\vec{B} = -\nabla V_m \quad ...(2)$$

where, $V_m = \mu_o I \Omega / 4\pi$

$$= \frac{\mu_o}{4\pi} \times \text{current} \times \text{solid angle} \quad ...(3)$$

and for collection of loops,

$$V_m = \frac{\mu_o}{4\pi} \sum_K I_k \ \Omega_k \quad ...(4)$$

V_m is a scalar potential whose negative gradient gives the magnetic induction $\vec{B}$. This is generally called a 'scalar potential' since it is scalar quantity, to distinguish it from the vector potential, which we shall soon introduced. It is clear that, in cases, in which $\vec{B}$ can be derived from a scalar potential, we must have curl $\vec{B} = 0$, since the curl of any gradient of scalar is zero.

Comment on Scalar Potential

The electrostatic potential is single valued. However, the potential V_m is no single-value since the solid angle is not single-valued. In face solid angle is not a single-valued function of position P. If point P makes a series of circuits through the loop, solid angle (Ω) will change to $\Omega + 4\pi n$ where n is some positive or negative integer. Unlike electrostatic potential, the difference between the scalar potential between any tow points is also undefined unless the path followed between the two points is specified. This is so because each time, the path links the circuit, the solid angle along the path changes by 4π and the difference in magnetic scalar potential changes by,

$$\frac{\mu_o}{4\pi} (\Omega + 4\pi - \Omega) = \mu_o I.$$

Thus, the magnetic scalar potential v_m is multiple valued. *This method may be used for calculating the potential of conductors of negligible cross-sections. It cannot be applied for points within the conductors themselves.* It can be used for points well outside the current circuit under consideration. We can choose the value of Ω and hence of V_m which goes to zero at infinity. Hence V_m can be defined by an equation similar to the electric potential with reference point chosen at infinity.

$$V_m = -\int_\infty^P \vec{B} . d\vec{l}$$

Unit

From relation $Vm = \mu_o I \ \Omega/4\pi$, the MKS unit has dimensions of μ_o times amperes. *The unit for V is not completely given.*

We have not interpreted V_m as representing energy for the reason that B does not directly represent a force on a real entity. If single magnetic pole really existed then $\vec{B}$ would be the force on unit pole at the point in question. Hence we say Vm as a sort of fictional potential energy per unit pole.

THE MAGNETIC DIPOLE

Suppose we have a current I flowing in a positive direction around a very small loop of area A, the normal to the loop being $\hat{n}$. If θ is the angle between $\hat{n}$ and the vector from the loop to a point P, as shown in Fig. 3.3, the solid angle subtended by A at P will be $A \cos \theta / r^2$, so that the scalar potential will be by eqn. (1)

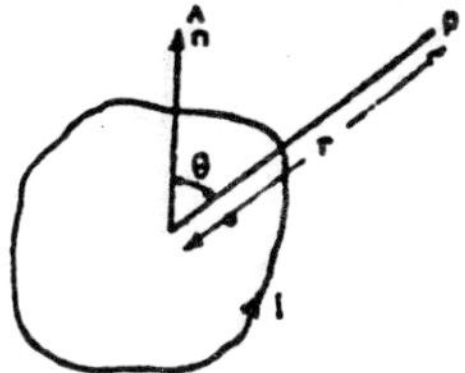

Fig. 3.3

$$\frac{\mu_0 I}{4\pi} \frac{A \cos \theta}{r^2}$$

This expression depends on position of the point P, just like the potential of an electric dipole,

$$V = \frac{1}{4\pi\varepsilon_0} \frac{P \cos}{r^2} = \frac{1}{4\pi\varepsilon_0} \frac{\vec{p}.\hat{r}}{r^2}$$

Thus, a small loop carrying a current has magnetic induction $\vec{B}$, whose lines of force are like those of an electric dipole, we call such a small loop a magnetic dipole and define the product IA as the dipole moment. If this moment is represented by m, we have,

$$\text{Potential of magnetic dipole} = \frac{\mu_0 m \cos \theta}{4\pi r^2} = \frac{\mu_0}{4\pi} \frac{\vec{m}.\hat{r}}{r^2}$$

$$= \frac{\mu_0}{4\pi} \frac{\vec{m}.\hat{r}}{r^3} \quad ...(2)$$

Magnetic Shell and its Equivalence with a Current Loop

In terms of this formulation, there is an interesting way in which we can regard the magnetic field of a current loop as carrying from a distribution of magnetic dipole over a surface spanning the loop shown in Fig. 3.4, we may divide the surface into may small loops, let a current I flow in each of them in positive direction. These currents will cancel each other on all the interior boundaries of the loops, but not over the outer boundary. Thus, the sum of the currents of all the small loops will give just the current originally present in the large loop. So that the magnetic field of the small loops must equal that of large loop. Each of the small loops of area A, however, produce a field like a dipole of moment IA. In other words, a distribution of dipoles over surface, whose dipole moment per unit area is I, distributed over any surface spanning the current loop will have the same magnetic induction at all points that the current loop itself has.

Strength of Magnetic Shell

It may be defined as the magnetic moment per unit area. Therefore, strength of the magnetic shell = IA/ A = I

Potential at a Point Very Close to the Surface

$$V_m = \frac{\mu_0 I}{4\pi} \times \text{solid angle}$$

in this case solid angle = 2π.

$$\text{Hence, } V_m = \frac{\mu_0 I}{4\pi} I\, 2\pi = \frac{\mu_0 I}{2} \quad ...(3)$$

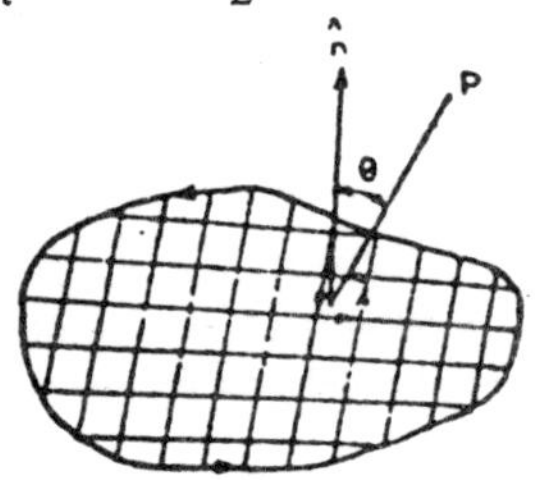

Fig. 3.4

At a Point Inside a Closed Magnetic Shell,

Solid angle = 4π

Hence, $V_m = \mu_o I$ ...(4)

AMPERE'S CIRCUITAL LAW OR AMPERE'S LAW

Ampere by using relation t = m B sing θ, in which t is torque, m is the magnitude of the magnetic moment and θ is the angle between the direction of B and normal to the wire, measured B by bringing a compass needle near a straight wire.

Even though m for a compass needle is not known, m may be regarded as constant independent of the position or orientation of the needle. Thus by measuring t and q one can obtain a relative measure of B for various distances r and for various currents I in the wire. The experimental results can be represented mathematically,

$$B \propto I/r.$$

We can convert this proportionality into an equality by inserting a proportionality constant $\mu_o/2\pi$ with the same reasoning as far Coulomb's law μ_o is permeability constant for free space

$$B = \frac{\mu_0 I}{2\pi r}$$

or $B\,(2\pi r) = \mu_o I$

The L.H.S. is $\oint \vec{B}.d\vec{l}$ for a path consisting of a circle of radius r centred on the wire. For all points on this circle $\vec{B}$ has the same constant magnitude of $\vec{B}$ and $d\vec{l}$ which is always tangent to the path of integration,

$$\oint \vec{B}.d\vec{l} = \oint B.dl = B\oint dl = B(2\pi r)$$

In this special case we can write the experimentally observed connection between $\vec{B}$ and I, $\oint \vec{B}.d\vec{l} = \mu_o I$ ...(5)

Statement of Ampere's Law

The law states that the line integral of $\vec{B}$ around a close path is equal to μ_o times the current enclosed by the path if the path does not enclose the current then,

$$\oint \vec{B}.d\vec{l} = 0.$$

PROOF OF AMPERE'S LAW–THE LINE INTEGRAL OF THE MAGNETIC FIELD VECTOR

We can prove the Ampere's law which is a fundamental relation between the current and the line integral of the magnetic field vector.

Line Integral for Irregular Path Which Encloses Current

Now we shall examine the line integral in the close path of integration around the current wire of an irregular shape as shown in Fig. 3.5(a). First determine the value of line integral along a small section dl of total path, which is shown in amplified form in Fig. 3.5(b). $\vec{B}$ along the are $d\vec{l}$ is given in the similar way as above.

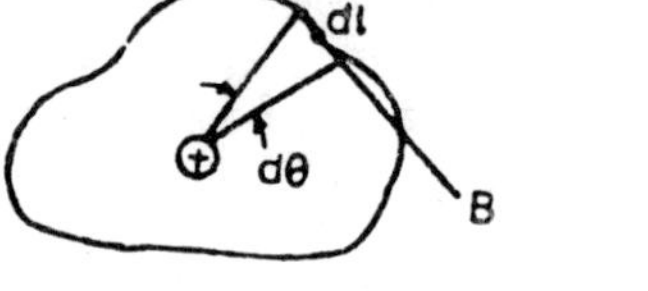

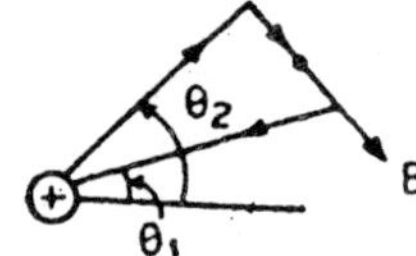

Fig. 3.5(a) Fig. 3.5(b)

$$\int_{are} \vec{B}.d\vec{l} = \frac{\mu_o I}{2\pi}\int_{\theta_1}^{\theta_2} d\theta = \frac{\mu_o I}{2\pi}(\theta_2 - \theta_1)$$

Along the radial path $\vec{B}$ and $d\vec{l}$ are perpendicular to each other where $\vec{B}.\ d\vec{l}$ is zero. So

$$\int_{\substack{\text{radial}\\ \text{path}}} \vec{B}.d\vec{l} = 0.$$

Clearly for the element of the path shown, therefore,

$$\int_{path} \vec{B}.d\vec{l} = \frac{\mu_o I}{2\pi}(\theta_2 - \theta_1)$$

Since the result is independent of the distance R, then by taking the summation of the contribution to the total integral by all segments into which the total path has been divided, it is clear that $\theta_2 - \theta_1 = 2\pi$ for closed path that encloses the current

$$\oint \vec{B}.d\vec{l} = \mu_o I.$$

If the path is taken either radial or parallel to the current, $\vec{B}$ and $d\vec{l}$ are normal to each other. Since any arbitrary $d\vec{l}$ can be analysed into radial, tangential and parallel components, it is clear that only the tangential component contributes to the integral. Therefore, one chooses an arbitrary path for the integration around the current. The path could again be divided into segments as above only tangential components will be considered and the same result.

Value of $\oint \vec{B}.d\vec{l}$ When Path does not Enclose Current

Here too the path which does not enclose the current is broken into segments as shown in Fig. 3.6(a). Based on ;the same consideration as in the case when path enclosed the current for the path shown in Fig. 3.6(b) two tangential contributions exist.

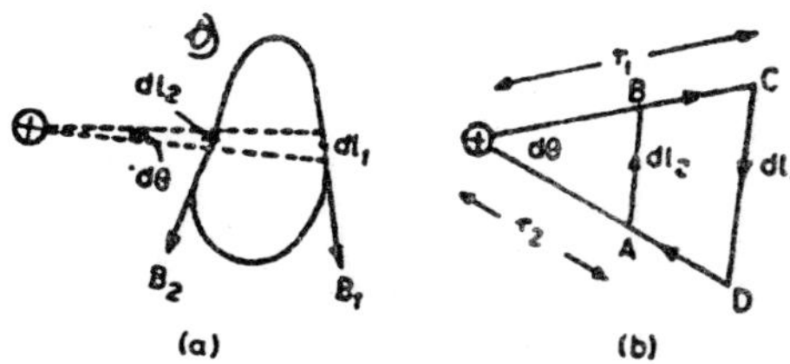

Fig. 3.6

Thus,
$$\int_{\substack{\text{total}\\\text{path}\\\text{ABCD}}} \vec{B}.d\vec{l} = \int_{\substack{\text{are}\\ dl_1}} \vec{B}.d\vec{l} + \int_{\substack{\text{are}\\ dl_2}} \vec{B}.d\vec{l}_2$$

But the two integrals on R.H.S. have equal and opposite values. However the value of integral vanishes for BC and AD, B and dl are perpendicular, hence $\vec{B}.\,d\vec{l} = 0$. For complete path, we have

$$\int_{ABCD} \vec{B}.d\vec{l} = \int_A^B \vec{B}.d\vec{l} + \int_B^C \vec{B}d\vec{l} + \int_C^D \vec{B}.d\vec{l} + \int_D^A \vec{B}.d\vec{l}$$

$$= \int_A^B \vec{B}.d\vec{l} + \int_C^D \vec{B}.d\vec{l}$$

$$= \int_A^B \vec{B}.d\vec{l}_2 - \int_D^C \vec{B}.d\vec{l}_1 \text{ (since } dl_2 \text{ and } dl_1 \text{ are opposite)}$$

B at a distance $r_1 = \dfrac{\mu_0 I}{2\pi r_1}$

B at a distance $r_2 = \dfrac{\mu_0 I}{2\pi r_2}$

Hence,
$$\int_{ABCD} \vec{B}.d\vec{l} = \frac{\mu_0 I}{2\pi r_2}\int_A^B dl_2 - \frac{\mu_0 I}{2\pi r_1}\int_D^C dl_1$$
$$= \frac{\mu_0 I}{2\pi r_2} AB - \frac{\mu_0 I}{2\pi r_1} DC$$

Now, from Fig. 4.20 (b),

$$DC/AB = r_1/r_2$$

or
$$DC = (r_1/r_2)\ AB$$

Putting this value of DC in the above equation, we get,

$$\int_{ABCD} \vec{B}.d\vec{l} = \frac{\mu_0 IAB}{2\pi r_2} - \frac{\mu_0 Ir_1 AB}{2\pi r_1 r_2} = 0$$

Thus, the results can be summarised,

$$\oint \vec{B}.dl = \mu_0 I \quad \text{if path links current} \qquad ...(6)$$
$$= 0 \quad \text{if path does not link current}$$

Ampere's Law and Gauss's Law of Electrostatics

Eqn. (6) is identical to Eqn. (3) in Chapter 4 for Gauss's law. Ampere's law is identical to Gauss's theorem of electrostatics for calculating magnetic field. As in electrostatics the differential form of Gauss's law is a fundamental equation of Maxwell electromagnetic theory. Similarly, Ampere's law will be also expressed as a fundamental equation of electromagnetic theory we will just see it. However, we can apply the Ampere's law where $\oint \vec{B}.d\vec{l}$ is easily evaluated *i.e.*, in problems having sufficient symmetry.

APPLICATIONS OF AMPERE'S LAW?

When do We Apply Ampere's Law

Like Gauss's law, Ampere's law is *always true* (for steady currents), but it is not always useful. The usefulness of Ampere's law depends on out ability to evaluate $\oint \vec{B}.d\vec{l}$ For this, we construct a closed path, called Amperian path near the current-distribution, for which the magnitude of field and angle between $\vec{B}$ and to the direction of path, have constant value; since then B-cos θ can be taken outside the integral sign. B is constant for uniform and symmetrical current-distribution and θ is constant for symmetrical path. *Thus, symmetry is crucial for the application of Ampere's law.* Therefore, we apply Ampere's law only when we have uniform and symmetrical, current-distribution and symmetrical Amperian path. The standard current-configurations which can be handled by Ampere's law are :

1. infinite straight lines
2. infinite planes
3. infinite solenoids
4. toroids.

When Ampere's law applies, it's by far the fasted method; when it doesn't, we'll have to fall back on the Biot-Savart law.

APPLICATIONS OF AMPERE'S LAW

We shall apply this law for calculating B in important cases given below:

1. $\vec{B}$ Near a Long Wire.

As we know that the line of magnetic induction for a long straight wire carrying a current I are concentric circles centred on the wire. The central dot in Fig. 3.7 represents a current I in the wire emerging from the page. The angle between $\vec{B}$ and $d\vec{l}$ is zero, hence $\vec{B}.d\vec{l}$ = BDl, then applying the Ampere's law We have, $\vec{B}$ at point P at a distance r

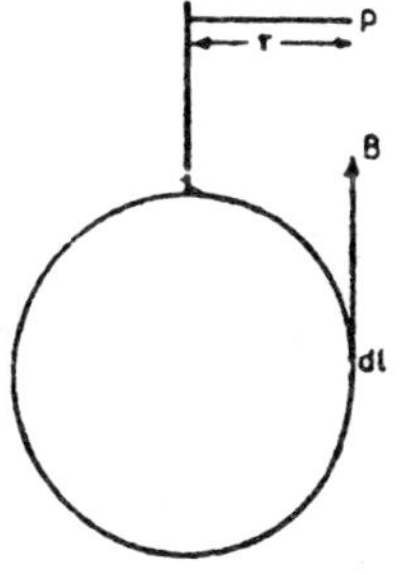

Fig. 3.7

$$\oint \vec{B}.d\vec{l} = \mu_o I$$

$$B.2\pi r = \mu_o I$$

$$B = \frac{\mu_0 I}{2\pi r} \quad ...(1)$$

In fact we may regard this as an experimental result consistent and also derivable from Ampere's law. This expression is same as obtained by Biot-Savart law.

It is interesting to compare Eqn. (1) with the expression for electric field near a long line of charge,

$$E = \frac{I}{2\pi\varepsilon_0} \frac{\lambda}{r}$$

Both are identical depending on constant multiple factor $\mu_0/2\pi$ and $1/2\,\pi\varepsilon_0$ and the factor responsible for the fields I and λ. Finally each field varies as $1/r$. They are derived in the same manner $\vec{E}$ by using Gauss's law and evaluating integral over Gaussian surface and $\vec{B}$ by using Ampere's law and evaluating for closed circular path. However, lines of $\vec{E}$ are perpendicular to the surface, while lines of $\vec{B}$ are everywhere tangent to the circular path.

2. Two Parallel Conductors

Whether we use Biot-Savart law or Ampere's law, B near a long wire at a distance r from the wire given by $\mu_0 I/2\pi r$. Hence, we can explain the force between the two currents in the same manner as discussed previously.

3. B for a Solenoid

A solenoid is a long wire wound in a close-packed helix and carrying a current I. For points near to a single turn, the behaviour of the wire is same as that of straight wire. The total field is the vector sum of the fields set up by all the turns. B is parallel to the solenoid axis at the interior points.

The field at external point is zero. For points such as P shown in Fig. 3.8(a) the field set up by the upper part of the solenoid turns marked O points to the left and tends to cancel the field set-up by the lower part of the solenoid turns mark ⊗, which points to the right as shown in Fig. 3.8(a).

Let us apply Ampere's law,

$$\oint \vec{B}.d\vec{l} = \mu_0 I$$

to the path abcd shown in Fig. 3.8(b). The total integral is sum of four integrals, one for each path of rectangle,

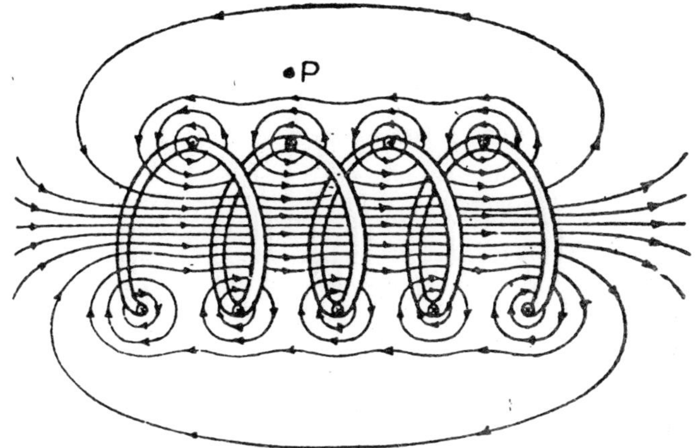

Fig. 3.8(a)

Fig. 3.8(b)

$$\int_{abcd} \vec{B}.d\vec{l} = \int_{a}^{d} \vec{B}.d\vec{l} + \int_{d}^{c} \vec{B}.d\vec{l} + \int_{c}^{b} \vec{B}.d\vec{l} + \int_{b}^{a} \vec{B}.d\vec{l}$$

for ba, B and dl are perpendicular to each other hence $\vec{B}.\ d\vec{l} = 0$ and similarly for dc $\vec{B}.\ d\vec{l} = 0$. The path cb is outside the solenoid $\vec{B}.\ d\vec{l}$ is zero because B is zero outside the solenoid. The only contribution to the field is due to the path ad thus for entire path,

$$\int_{abcd} \vec{B}.d\vec{l} = \int_{a}^{d} \vec{B}.d\vec{l}$$

B and dl are parallel inside, hence,

$$\int_{abcd} \vec{B}.d\vec{l} = B\int_{a}^{d} dl, \text{ since } \int_{a}^{d} dl = \text{lenght ad} = h$$

$$= Bh$$

The net current I that passes through the area bounded by the path of integration is not the same as the current I_0 in the solenoid because path of integration encloses more than one turn.

Let n be the number of turns per unit length, then

$$I = J_0(nh)$$

So, $$Bh = \mu_0 I_0 nh \quad ...(2)$$

The relation is obtained for an infinitely long ideal solenoid, it holds for actual solenoid for internal points near the centre of the solenoid. The relation does not depend on the diameter or the length of the solenoid and B is constant over the solenoid cross section. If N are the total number of turns and l is the length of wire of the solenoid, then

$$n = N/l$$

$$B = \mu_0 I_0 (N/l) \quad ...(3)$$

same expression as derived by Biot-Savart law.

4. A Toroid

Fig. 3.9 shows a toroid. A toroid may be visualised as a doughtnut with wire would uniformly around the doughnut. The toroid may be considered as a solenoid that has been bent into a circle with the ends joined. Since the axis of the resulting bent solenoid is circular, the lines of $\vec{B}$ are also circular.

Let us apply Ampere's law to the circular path of integration of radius r,

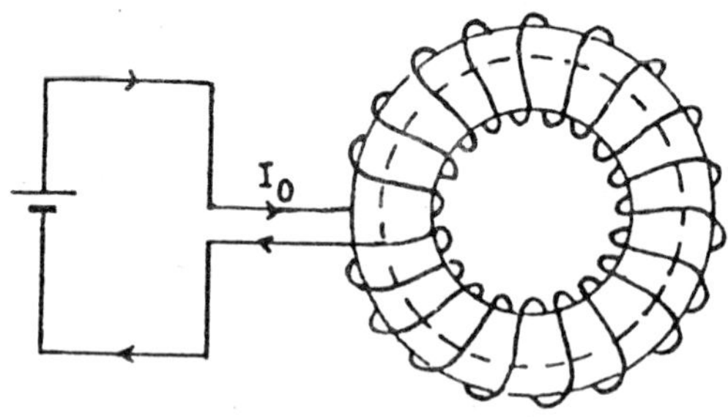

Fig. 3.9

$$\oint \vec{B}.d\vec{l} = \mu_0 I$$

$$B.2\pi r = \mu_0 I_0 N$$

where I_0 is the current in the toroid windings and N are the total number of turns. Therefore,

$$B = \frac{\mu_0}{2\pi} \frac{I_0 N}{r} \quad ...(4)$$

In contrast to the solenoid, B is not constant over the cross-section of a toroid.

For External Point

As discussed in the case of solenoid the field at outside point is zero. So here also B is zero because toroid is a bent circular solenoid.

CHARACTER OF $\vec{B}$ LINES AND THE DIVERGENCE OF $\vec{B}$

In our study of electrostatics, we first found simple methods of obtaining the field by integrating the contribution resulting from all points charges according to Coulomb' law. We soon found, however that this was method of restricted usefulness. The really powerful tools of electrostatics come only when we apply general analytical methods, introducing the potential, Gauss's theorem, Poisson's equation, and similar relations. Here too, we shall introduce general analytical methods. We shall first show the div. $\vec{B}$ is always zero.

The lines of magnetic field are closed circle around the axis of a current element. In other words, the lines of magnetic field produced by currents always close on themselves. This means of course, that if an infinitesimal region of space were examined, the total number of $\vec{B}$ lines entering the specified region must be equal to the total number of $\vec{B}$ lines leaving the specified region. As we have already established for steady current where the continuity prevails for the number of lines of electric currents is characterised mathematically by div. $\vec{J} = 0$. Same condition is satisfied by vector $\vec{B}$ as $\vec{B}$ lines are continuous, hence we must have div. $\vec{B} = 0$.

To be more specific, since the magnetic field lines do not have sources in contrast to the case of electric field where the lines emerge from positive charges and end on the negative charges, as we know the divergence of a vector signifies the starting or stopping of the vector, we should expect that div.

$$\vec{B} = 0 \quad ...(1)$$

Proof:

We can prove $\text{div}.\vec{B} = 0$ from Biot-Savart law and using vector analysis. Noting that for small element of current $I d\vec{l} = \vec{I} dl$. According to Biot-Savart law, the field of complete circuit is given by,

$$\vec{B} = \frac{\mu_0}{4\pi} \int \frac{\vec{I}\, dl \times \hat{r}}{r^2}.$$

taking the divergence of both sides,

$$\text{div}\ .\ \vec{B} = \nabla.\vec{B} = \nabla.\left(\frac{\mu_0}{4\pi} \int \frac{\vec{I}\, dl \times \hat{r}}{r^2}\right)$$

$$= \frac{\mu_0}{4\pi} \int \nabla\left(\frac{\vec{I} \times \hat{r}}{r^2}\right) dl$$

Since, it makes no difference whether we first evaluate integral then take divergence or take the divergence first then evaluate the integral.

therefore $$\text{div}.\ \vec{B} = \frac{\mu_0}{4\pi} \int \nabla.\left(\frac{\vec{I} \times \hat{r}}{r^3}\right) dl$$

Since $$\nabla\left(\frac{I}{r}\right) = -\ 1/r^2$$

hence $$\text{div}.\ \vec{B} = -\ \frac{\mu_0}{4\pi} \int \nabla.\left[I \times \nabla\left(\frac{1}{r}\right)\right] dl$$

But according to the vector analysis expansion,

$$\nabla.(\vec{A} \times \vec{B}) = \vec{B}.\nabla \times \vec{A} - \vec{A}.\nabla \times \vec{B}]$$

therefore, $$\text{div}.\ \vec{B} = -\ \frac{\mu_0}{4\pi} \int \left[\nabla\left(\frac{I}{r}\right).\nabla \times \vec{I} - \vec{I}.\nabla \times \nabla\left(\frac{I}{r}\right)\right] dl$$

To interpret this result it is recalled that $\vec{B}$ is being specified at the field point, where as the current is that at the source points specified by dl. Say the field point is specified by the variable. (x, y, z) the source point might be specified by the variable (x', y', z') But since the divergence operation is being taken with respect to the field point variables which are independent of the source point variables, then

$$\nabla\left(\frac{I}{r}\right).\nabla \times \vec{I} = 0.$$

The second term, which is the curl of the gradient of scalar is identically zero, then it follows, that always div. $\vec{B} = 0$.

AMPERE'S LAW IN CURL FORM OR CURL $\vec{B}$ AND STOKES' THEOREM

The Ampere circuital law is a simple relation between magnetic vector $\vec{B}$ and current I. The law is easily applied to calculate $\vec{B}$ when sufficient symmetry exists. *If one is interested to apply the law the problems of a general nature without a high degree of symmetry, it is necessary to relate the magnetic vector* $\vec{B}$*with the current density.* The situation here is rather like that div. $\vec{E} = \rho/t_o$, which relates the divergence of $\vec{E}$ with the charge density to specify the conditions at a point. Ampere's law can be written by using the definition of I and current density,

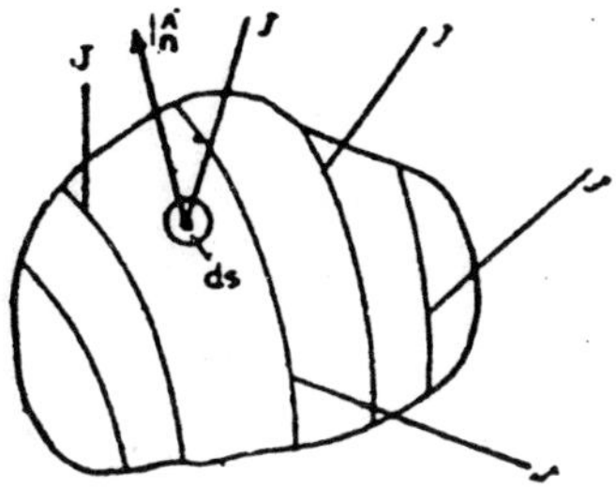

Fig. 3.10

$$\oint \vec{B}.d\vec{l} = \mu_o I$$

$$= \mu_o \int_s \vec{J}.\hat{n}ds \qquad ...(1)$$

where the area s is that specifying the region through which the current exists, $\hat{n}$ is a unit vector normal to elementary area ds in Fig. 3.10. Applying the Stokes' theorem,

$$\oint \vec{B}.d\vec{l} = \int_s (\nabla \times \vec{B}).\hat{n}ds \qquad ...(2)$$

Comparing Eqn. (1) and (2), we get,

$$\text{Curl } \vec{B} = \tilde{N} \times \vec{B} = \mu_o \vec{J} \qquad ...(3)$$

Eqn. (3) is the point statement of Ampere's law corresponding to the integral statement, taking the two corresponding forms are,

$$\left.\begin{aligned}\oint \vec{B}.d\vec{l} &= \mu_0 I\\ \nabla\times\vec{B} &= \mu_0 J\end{aligned}\right\} \qquad \text{and} \qquad ...(4)$$

There expressions bear the same general relationship to each other that Gauss's law in electric field,

$$\oint \vec{E}.d\vec{s} = \frac{Q}{\varepsilon_0}$$

or, $\quad \text{div. } \vec{E} = \rho/\varepsilon_0$

It is interesting to write down,

$$\left.\begin{aligned} &\text{div. } \vec{B} = \nabla.\vec{B} = 0\\ \text{and} \quad &\text{Curl } \vec{B} = \nabla\times\vec{B} = \mu_0 J\\ \text{in electrostatic,}\\ &\text{div. } \vec{E} = \nabla.\vec{E} = \rho/\varepsilon_0\\ \text{and} \quad &\text{Curl } \vec{E} = \nabla\times\vec{E} = 0\end{aligned}\right\} \qquad ...(5)$$

A Further Comment on Scalar Potential

We see from Eqn. (4), $\nabla \times \vec{B} = \mu_0 J$, that in a region containing no current flow curl $\vec{B} = 0$, but this does not hold within a conductor. In empty space where curl $\vec{B} = 0$, we can introduce as we have done a scalar potential but this cannot be done inside a conductor carrying a current. The scalar potential we introduced for a single loop of current was multiple-valued, that is as we integrated $\vec{B}$ around a contour enclosing the current, the potential increased by $\mu_0 I$, so that if we integrated n times around the contour we, should increase the potential by n $\mu_0 I$. In other words, for any single value of potential, all values obtained by adding or subtracting integral multiples of $\mu_0 I$ are equally legitimate. If now, we have many current loops, in our space, we can change the potential by integral multiples of μ_0 times of any one of the currents, by traversing suitable contours, with enough current loops, this means that we can obtain almost any value of the potential we desire, at a given point of space, if the current is distributed over the volume, so that we can enclose any amount by traversing a suitable contour within the volume, we can obtain any value of potential whatever at a given point of space. In this case the usefulness of the idea of potential breaks down completely. The integral of $\vec{B}$ is not independent of path, and there is no unique way of defining a scalar potential.

THE VECTOR POTENTIAL

It is clear from our discussion of the preceding section that a scalar potential is not useful for discussing a magnetic field, except in the special case where the current flows in a single loop. Mathematically, we have that curl $\vec{B}$ is not in general zero, so that the conditions for introducing a scalar potential do not exist. *There is, however, another quite different way of setting up a potential function, which is much more general, and equally useful. This is to set up what is called a 'vector potential'*. The fundamental characteristic of a potential is that it is a function that we differentiate to get the vector field. The reason why the potential is a useful device in electrostatics is that it is easy to compute it from charge distribution, by the solution of Poisson's equation.

The vector potential solution for the magnetic field of currents has the same useful features. The existence of vector potential is based on div. $\vec{B} = 0$, always satisfied by magnetic induction. By a well-known theorem of vector analysis, div. curl $\vec{A} = 0$, where $\vec{A}$ is any arbitrary vector, that is, the divergence of any curl is zero. It seems reasonable to assume from this that we can write.

$$\vec{B} = \text{curl } \vec{A} \qquad \text{...(1)}$$

where $\vec{A}$ is a vector function of position, which is the vector potential. In other words, the vector $\vec{A}$ the curl of which is equal to the magnetic induction $\vec{B}$ is known as vector potential.

Eqn. (1) can be written as,

$$B_z\hat{k} = \left(\frac{\delta A_z}{\delta y} - \frac{\delta A_y}{\delta z}\right)\hat{i} + \left(\frac{\delta A_x}{\delta z} - \frac{\delta A_z}{\delta x}\right)\hat{j} + \left(\frac{\delta A_y}{\delta x} - \frac{\delta A_x}{\delta y}\right)$$

Now equating the components on both sides of the equation,

$$\left.\begin{aligned} \vec{B}_x = (\nabla \times \vec{A})_x &= \left(\frac{\delta A_z}{\delta y} - \frac{\delta A_y}{\delta z}\right) \\ \vec{B}_y = (\nabla \times \vec{A})_y &= \left(\frac{\delta A_x}{\delta_z} - \frac{\delta A_z}{\delta x}\right) \\ \vec{B}_z = (\nabla \times \vec{A})_z &= \left(\frac{\delta A_y}{\delta x} - \frac{\delta A_x}{\delta y}\right) \end{aligned}\right\} \qquad \text{...(2)}$$

where A_x, A_y, and A_z are the corresponding components of the vector potential A.

Divergence of $\vec{A}$

We shall consider in next section that div. $\vec{A} = 0$.

Let us prove it. As div.$\vec{B} = 0$, we have written,

$$\vec{B} = \nabla \times \vec{A}, \text{ it guarantees that,}$$

$$\nabla.B = \nabla.\,(\nabla \times \vec{A}) = 0$$

The field A is called the vector potential. In cases where $\nabla \times \vec{B} = 0$, we have $\vec{B} = -\nabla V_m$, where V_m is scalar potential. As we have discussed that the Scalar potential V_m was not completely specified by its definition. V_m is a multivalued function. If we found V_m for some problem, we can always find another potential that is equally good by adding a constant,

$$V_m' = V_m + C$$

The new potential V_m' gives the same magnetic field. So we can say that corresponding to a given value of magnetic field, there can be two or more scalar potentials differing by addition of a constant C, since gradient ∇ C is zero we have V'_m and V_m represent the same phenomena i.e, the same magnetic field.

Similarly, we can have different vector potentials $\vec{A}$ which give the same magnetic field. Again, because $\vec{B}$ is obtained from $\vec{A}$ by differentiating as given in Eqn. (2) adding a constant to A does not change anything physical. But there is even more latitude for A. We can add to A any field which is the gradient of some scalar field, without changing to physical phenomena.

Suppose we have a vector potential A that gives correctly the magnetic field $\vec{B}$ and the other new vector potential A' will give the same field $\vec{B}$ if they have the same curl *i.e.*,

$$\vec{B} = \nabla \times \vec{A} = \nabla \times \vec{A}'.$$

Therefore, $\nabla \times \vec{A} - \nabla \times \vec{A} = \nabla \times (\vec{A}' - \vec{A}) = 0$

But if the curl of a vector is zero it must be the gradient of some scalar field, say ϕ. So,

$$A' - A = \nabla \phi$$

$$A' = A + \nabla \phi \qquad ...(3)$$

will be an equally satisfactory vector potential, leading to the same field $\vec{B}$.

It is usually convenient to vary $\vec{A}$ by imposing some condition on it (in the same way that we found it convenient to choose the potential V_m zero at large distances.) We can, for instance, restrict $\vec{A}$ by choosing arbitrarily value of divergence of $\vec{A}$, we can always do without affecting $\vec{B}$. This is because although A' and A have the same curl, and give the same $\vec{B}$, they do not need to have the same divergence. In fact,

$$\nabla . \vec{A}' = \nabla . \vec{A} + \nabla^2 \phi$$

And by a suitable Choice of ϕ, we can make $\nabla . \vec{A}'$ any thing we wish.

What should be the choice of $\nabla . \vec{A}$? The choice should be made to get the greatest mathematical convenience and will depend on the individual problem. For magnetostatics, we will make the simple choice,

$$\nabla . \vec{A} = 0$$

Then the complete definition of $\vec{A}$ is for magnetostatics.

$$\nabla \times \vec{A} = \vec{B}$$

and

$$\nabla . \vec{A} = 0 \qquad ...(4)$$

THE VECTOR POTENTIALS OF KNOWN CURRENTS

As we know that the magnetic field vector $\vec{B}$ is determined by the currents, the vector potential $\vec{A}$ is related to magnetic field B by the relation,

$$\vec{B} = \text{curl } \vec{A},$$

So $\vec{A}$ must also be determined by the currents. We want to find $\vec{A}$ in terms of the currents. We begin with the basic equation,

$$\nabla \times \vec{B} = \mu_o \vec{J},.$$

Substituting $\vec{B}$ in terms of vector $\vec{A}$, the equation reduces, to

$$\nabla \times \nabla \times \vec{A} = \mu_o \vec{J}$$

using the vector identity.

$$\nabla \times \nabla \times A = \nabla (\nabla . \vec{A}) - \nabla^2 A = \mu_o J$$

Now since ∇. A = 0, therefore,

$$\nabla^2 \vec{A} = \mu_o \vec{J}$$

$$\left.\begin{array}{l} \nabla^2 \vec{A} = \mu_o \vec{J} \\ \text{or in terms of components,} \\ \nabla^2 A_x = -\mu_o J_x \\ \nabla^2 A_y = -\mu_o J_y \\ \text{and} \\ \nabla^2 A_z = -\mu_o J_z \end{array}\right\} \quad ...(1)$$

Each of the equation given in eq. (1) is mathematically identical to the equation of electrostatics *i.e.*, Poisson's equation,

$$\nabla^2 \vec{A} = -\rho/\varepsilon_o$$

Now we know that the solution of Poisson's equation gives the value of V(x, y,z), hence a very simple process for calculating electrostatics field of a charge distribution,

$$V(x_1, y_1, z_1) = \frac{1}{4\pi\varepsilon_o} \int_V \frac{\rho(x_2, y_2, z_2)}{r_{12}} dv_2$$

where r^{12} is the separation between the points (x_2, y_2, z_2) and (x_1, y_1, z_1). So we see immediately from an analogy the general solution for equation,

$$\nabla^2 A_x = -\mu_o J_x$$

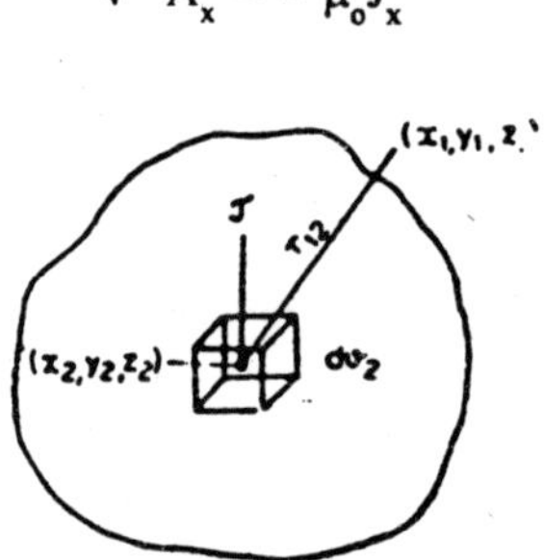

Fig. 3.11

As, $$A_x = \frac{\mu_o}{4\pi}\int_V \frac{J_x(x_2, y_2, z_2)}{r_{12}} dv_2$$

i.e., simply replacing ρ/ε_o by $\mu_o J_x$.

The dv_2 and r_{12} are shown in Fig. 4.26. Similarly we have for A_y and A_z. And when three solutions are combined, we get the solution of

$\nabla^2 A = \mu_o \vec{J}$, *i.e.*,

$$A(x_1, y_1, z_1) = \frac{\mu_o}{4\pi}\int_V \frac{J_x(x_2, y_2, z_2)}{r_{12}} dv_2 \qquad ...(2)$$

One can verify by direct differentiation that this integral satisfies condition $\nabla \cdot \vec{A} = 0$, so long as $\nabla \cdot \vec{J} = 0$, which must happen for steady currents. So by using Eqn. (2), we can find vector potential $\vec{A}$ due to the current density $\vec{J}$.

If the current is flowing in a wire having diameter very small as compared with the dimension, equation (2) can be modified for a thin wire of area of cross-section S and length dl, dv = Sdl,

hence, $$Jdv = JSdl$$

as current density is defined as current per unit area, dS = I, hence

$$Jdv = Idl$$

Putting in equation 2, we have,

$$A = \frac{\mu_o}{4\pi}\oint \frac{Idl}{r_{12}} \qquad ...(3)$$

where r_{12} is the distance of dl to the point at which we are calculating potential and the integral is taken around a whole curcuit.

APPLICATIONS OF VECTOR POTENTIALS

As, an example, we shall calculate the magnetic field by using vector potential method.

1. The B of Straight Wire

Let us consider a long straight wire of radius a carrying a steady current I which is uniformly distributed throughout its cross-section and the wire is placed in coordinate system as shown in

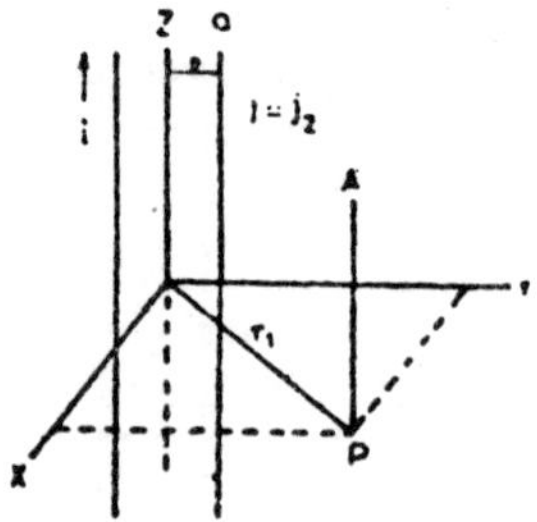

Fig. 3.12

Fig. 3.12 along the z-axis. The current density vector $\vec{J}$ will have, in this case, only z-component. It s magnitude is given by,

$$J_z = I/\pi a^2$$

inside the wire and is zero outside. Since $J_x = 0$ and $J_y = 0$, we immediately have,

$$A_x = 0, \; A_y = 0$$

To get A_z we can use our solution for the electrostatics potential V of a wire with a uniform charge density.

For points outside, an infinite charge cylinder, the electrostatic potential is,

$$V = \frac{\lambda}{2\pi\varepsilon_o} \log r'$$

where $r' = \sqrt{x^2 + y^2}$ and λ is the charge per unit length and is equal to the charge contained in a unit length of the cylinder, *i.e.*,

$$\lambda = \pi a^2 \rho$$

therefore,

$$V = \frac{I}{2\pi\varepsilon_o} \pi a^2 \rho \log r'$$

The corresponding vector potential Az is obtained by replacing ρ/ε_o by $\mu_o J_z$, *i.e.*,

$$A_z = -\frac{I}{2\pi} \pi a^2 \mu_o J_z \log r'$$

putting $I = \pi a^2 J_z$,

$$A_z = -\frac{\mu_o I}{2\pi} \log r'$$

Now, we can find B by using Eqn. (1),

$$B_x = \left(\frac{\delta A_z}{\delta_y} - \frac{\delta A_y}{\delta_z} \right)$$

Putting A_z given above and noting that $A_y = 0$, we get,

$$B_x = -\frac{\mu_0 I}{2\pi}\frac{\delta}{\delta y}(\log r'),\ r' = \sqrt{x^2 + y^2}$$

After simplification,

$$B_x = -\frac{\mu_0 I}{2\pi}\frac{y}{r'^2}$$

Similarly, $$B_y = \left(\frac{\delta A_x}{\delta_z} - \frac{\delta A_z}{\delta_x} \right)$$

as $A_x = 0$, and substituting A_z

$$= \frac{\mu_0 I}{2\pi}\frac{\delta}{\delta x}(\log r') = \frac{\mu_0 I}{2\pi}\frac{x}{r'^2}$$

and, $$B_z = \left(\frac{\delta A_y}{\delta x} - \frac{\delta A_x}{\delta y} \right) = 0, \text{ since } A_x = A_y = 0.$$

The magnitude of the magnetic field B is given by,

$$B = (B_x^2 + B_y^2 + b_z^2)^{1/2}$$

Substituting the values obtained above, we get

$$B = \frac{\mu_0 I}{2\pi r'^2}(x^2 + y^2)^{1/2}$$

$$= \frac{\mu_0 I}{2\pi r'} \qquad ...(2)$$

This is the same result as obtained earlier by Biot-Savart law and Ampere's circuital law.

2. The Field of a Small Current Loop-The Magnetic Dipole

Let us use the vector potential method to find the magnetic field of a small loop of current. By small loop, we mean simply that we are interested in the fields only at large distances compared with the size of the loop. Such a loop produces magnetic field like the electric field from an electrical dipole. It will be shown below that any small loop is a magnetic dipole.

Consider a rectangular loop and choose out coordinates as shown in Fig. 3.13(a). There are no currents in the z-direction so A_z is zero. There are currents in the x-direction on the two-sides of length a. In each leg, the current density (and current) is uniform. So the solution for A_x is just like the electrostatic potential from two charged rods s shown in Fig. 3.13(b). Since the rods

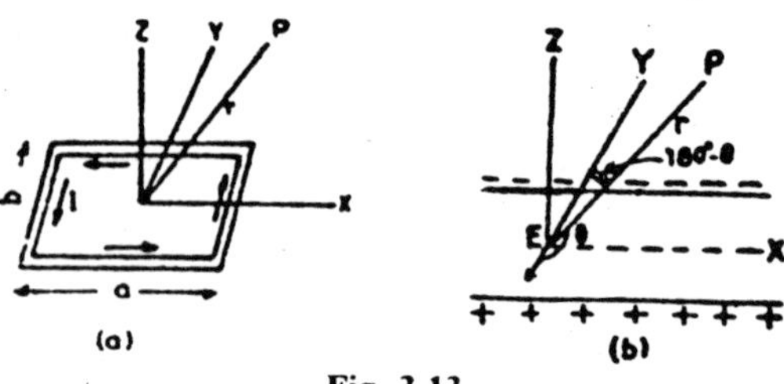

Fig. 3.13

have opposite charges, the electric potential at large distances would be just the dipole potential at the point P, the potential would be,

$$V = \frac{I}{4\pi\varepsilon_0} \frac{\vec{p}.\hat{r}}{r^2}$$

where P is the dipole moment of the charge distribution.

The dipole moment in the present case is the total charge on one rod times the separation between them.

$$P = \lambda \text{ a b}$$

where λ is the linear density of charge on the rods.

If α is the are of cross-section of the wire, ρ is the charge per-unit volume, them

$$\lambda = \text{volume of unit length of wire} \times \rho$$
$$= \alpha \rho$$

hence,

$$P = \alpha \rho \text{ ab}$$

and

$$V = \frac{1}{4\pi\varepsilon_0} \frac{P\cos\theta}{r^2}$$

where θ is the angle between the direction of $\vec{p}$ and $\vec{r}$ direction of $\vec{p}$ is same as $\vec{E}$ with negative y-direction,

$$\cos(180 - \theta) = y/r$$

$$\cos\theta = -\,y/r$$

therefore, $$V = -\frac{1}{4\pi\varepsilon_o}\frac{P}{r^2}\frac{y}{r}$$

putting the value of P given above,

$$V = -\frac{1}{4\pi\varepsilon_o}\frac{\alpha\rho ab}{r^2}\frac{y}{r}$$

From this A_x is directly written by replacement of ρ/ε_o by $\mu_o J$ *i.e.*,

$$A_x = -\frac{1}{4\pi}\frac{\mu_o J\alpha\, ab}{r^2}\frac{y}{r}$$

$$J\,\alpha = \text{current } I$$

hence, $$A_x = -\frac{\mu_o I\, ab}{4\pi}\frac{y}{r^3} \qquad ...(3)$$

Similarly, A_y can be calculated from the electric potential due to two oppositely charged rod lying parallel to y-axis separated by distance α.

$$V = \frac{1}{4\pi\varepsilon_o}\frac{P\cos\theta}{r^2}$$

Since $$\cos\theta = x/r,$$

$$V = \frac{1}{4\pi \in b}\frac{\rho\alpha\, ab}{r^2}\frac{x}{r}$$

hence A_y is obtained by replacing ρ/ε_o by $\mu_o J$

$$A = \frac{\mu_o I\, ab}{4\pi}\frac{x}{r^3} \qquad ...(4)$$

Therefore, the vector potential

$$A = (A_x^2 + A_y^2 + A_z^2)^{1/2}.$$

Putting A_x and A_y obtained above and also noting that Az = 0,

we get, $$A = \frac{\mu_o I\, ab}{4\pi}\frac{\left(x^2+y^2\right)^{1/2}}{r^3}$$

but $$r^2 = x^2 + y^2 + z^2$$

hence $$x^2 + y^2 = r^2 - z^2$$

or $$(x^2 + y^2)^{1/2} = r\sqrt{1 - x^2/r^2}\; z/r = \cos\theta_z$$

hence $$A = \frac{\mu_o I ab}{4\pi}\frac{1}{r^2}\sqrt{1-\cos^2\theta_z}$$

where θ_z is the angle which r makes with z-axis,

$$A = \frac{\mu_o}{4\pi} I\,ab \frac{\sin\theta_z}{r^2}$$

The quantity Iab which determines the strength of A and is product of area of the loop and current. This product is known as m, the magnetic dipole moment. Hence,

$$A = \frac{\mu_o}{4\pi}\frac{m\sin\theta_z}{r^2}$$

$$= \frac{\mu_o}{4\pi}\frac{\vec{m}\times\hat{r}}{r^2} \quad ...(5)$$

This will be the same at every point on a circle around the dipole. $\overline{A}$ is perpendicular to the plane containing the dipole and point of observation .

The magnetic field $\vec{B}$ can be calculated from Eqn. (6), as,

$$B_x = \left(\frac{\partial A_z}{\partial y} - \frac{\partial A_y}{\partial z}\right)$$

$$= -\frac{\delta A_y}{\delta_z}, \text{ since } A_z = 0.$$

Substituting A_y from Eq. (7), B_x reduces as given below.

$$B_x = -\frac{\partial}{\partial_z}\left(\frac{\mu_o m}{4\pi}\cdot\frac{x}{r^3}\right) \text{ where } m = Iab \text{ and } r = (x^2 + y^2 + z^2)^{1/2}$$

Substituting this, B_x reduces to

$$B_x = \frac{\mu_o m}{4\pi}\cdot\frac{3xz}{r^5}$$

Similarly,

$$B_y = \frac{\mu_o m}{4\pi}\cdot\frac{3yz}{r^5}$$

The value B_z is given in Eqn. (7), *i.e.*,

$$B_z = \frac{\partial A_y}{\partial x} - \frac{\partial A_x}{\partial y}$$

Substituting the values from Eqn. (7) and Eqn. (7a),

$$B_z = \frac{\mu_0 m}{4\pi}\left[\frac{\partial}{\partial x}\left(\frac{x}{r^2}\right) + \frac{\partial}{\partial y}\left(\frac{y}{r^3}\right)\right]$$

$$= \frac{\mu_0 m}{4\pi}\left[\frac{2}{r^3} - \frac{3}{r^5}\left(x^2 + y^2\right)\right]$$

Since, $r^2 = x^2 + y^2 + z^2$

after substitution, we get

$$B_z = -\frac{\mu_0 m}{4\pi}\left(\frac{1}{r^3} - \frac{3z^2}{r^5}\right)$$

therefore, the magnitude of B,

$$B = [B_x^2 + B_y^2 + B_z^2]^{1/2}$$

On substituting the values obtained above, we get,

$$B = \frac{\mu_0 m}{4\pi r^3}\sqrt{1 = 4\cos^2\theta} \qquad ...(8)$$

This expression is same as that of the electric dipole oriented along the z-axis. The current loop is, therefore, called a magnetic dipole.

3. Change in Magnetic Field at a Current Sheet

Consider a thin sheet of copper of uniform thickness t located in the xz plane in which a current flows with constant density and direction everywhere within the metal. Let the current flows is the x-direction. Fig. 3.14 shows a portion of the sheet of infinite extent. If the current density inside the metal is j in amp/metre^2 then every metre of height *i.e.*, of sheet in z-direction have an area t metre^2 and will include a ribbon of current amounting to jt ampere. We call this the surface current density or simply sheet current density and represented by j_8, j_8 is a useful quantity and determines the change in the magnetic field from one side of the sheet to the other.

The total magnetic field is the sum of the two fields, one due to sheet and the other from some external source present, there, Consider a rectangular path abcd, ab side is in front of the surface and dc behind with the short sides bc and da piercing the sheet. Let B_z + sheet and B_z – be that just behind it the line integral of B around abcd, is given by,

$$\int_{abcd} \vec{B}.d\vec{l} = \int_{ab} \vec{B}.d\vec{l} + \int_{bc} \vec{B}.d\vec{l} + \int_{cd} \vec{B}.d\vec{l} + \int_{dc} \vec{B}.d\vec{l}$$

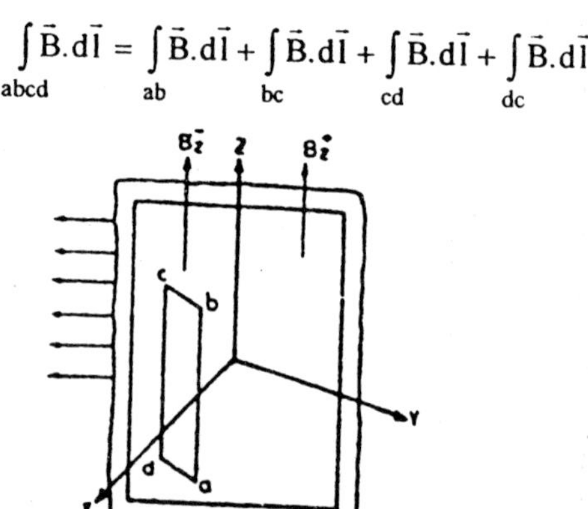

Fig. 3.14

Since we have considered the thin sheet and the point just near the sheet, hence bc and da sides are much shorter than ab and cd. Hence their contribution is neglected. Therefore,

$$\int_{abcd} \vec{B}.d\vec{l} = \int_{ab} \vec{B}.d\vec{l} + \int_{cd} \vec{B}.d\vec{l}$$

$$= (B_z^+ - B_z^-).l \quad ab = cd = l$$

Fields in front of the sheet and just behind it are just opposite to each other. Now using the Ampere's law

$$\int_{abcd} \vec{B}.d\vec{l} = \mu_o \times \text{current enclosed by rectangle}$$

$$= \mu_o\, j_s l$$

or, $$(B_z^+ - B_z^{-1})\, l = \mu_o\, j_s l$$

or, $$B_z^+ - B_z^- = \mu_o\, j_s \qquad ...(9)$$

A current sheet of density j_s increases the component of $\vec{B}$ which is parallel to the surface and perpendicular to j_s. This increase is the same as the change in electric field at a sheet of charge.

If the sheet is the only current source (the other external source is contributing). Then, of course, the field is symmetrical about the sheet. We have,

$$B_z + = \frac{1}{2}\mu_o j_8$$

$$\text{and } B_z - = -\frac{1}{2}\mu_o j_8$$

Suppose there are two sheets carrying equal and opposite surface currents as shown in Fig. 3.15. with on other sources around. The direction of current flow is perpendicular to the plane of the paper, out on the left and in on the right. The field between the sheet is and outside the sheets is zero

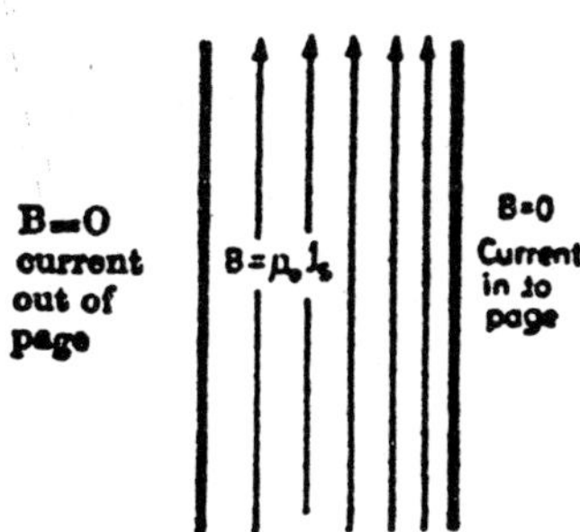

Fig. 3.15

$$\mu_o j_z = \left(\frac{1}{2}\mu_o j_8 + \frac{1}{2}\mu_o j_8\right)$$

$$\frac{1}{2}\mu_o j_8 + \frac{1}{2}\mu_o j_8 = 0$$

ASTON'S MASS SPECTROGRAPH

It is an improvement on Thomson method to increase the accuracy by modifying the principle of operation.

Principle

The main modification introduced is that the electric and magnetic fields are not applied simultaneously and in parallel direction as Thomson did, but successively, first the electric field and then the magnetic field.

Their directions being at right angles to each other. Therefore, the electric field produces a dispersion of the positive rays with respect to the velocity, while the magnetic field brings to a focus the dispersed rays.

Apparatus and its Working

A narrow beam of positive ions is made to pass through two parallel narrow slits A and B (0.01 mm in radius and 20 cm. apart) and a very fine beam is obtained. This beam is subjected to a uniform electric field intensity maintained between the plates C and D. The beam as a whole is deflected towards the negative electrode D. The beam is also dispersed through a certain angle because the beam contains ions of different velocities. The fast moving ions remain for a shorter time under the influence of electric field will be deflected less than the slower ones taking a longer time to traverse the field. Thus the beam is spread out into an electric spectrum according to different velocities. The divergent beam limited by slit S next enters the magnetic field B⊥E and the beam is bent in a direction opposite to that caused by E. Again the slowly moving ions are deflected more than the faster ions. Thus the beam can be focused at some point on photographic plate, if S is the linear displacement of the ions and l the length of their path in the electric field, then from Eq. (1),

$$S = \frac{eEl^2}{2mv^2}$$

(Force eE = ma, a = eE/m, initial velocity is zero

$$\left. S = \frac{1}{2}at^2, \; t = \frac{l}{v}, \; S = \frac{1}{2}\frac{eE}{m}\frac{l^2}{v^2} \right)$$

Similarly if S' is linear displacement and l' the length of path of the ion in the magnetic field

$$S' = \frac{1}{2}\frac{eB}{mv}l'^2$$

$$\left(\text{Force} = evB = ma', \; a' = \frac{evB}{m}, \; S' = \frac{1}{2}a'\,t'^2, \; t' = \frac{l'}{v} \right.$$

$$\left. S' = \frac{1}{2}\frac{evB}{m}\frac{l'^2}{v^2} = \frac{1}{2}\frac{eB}{mv}l'^2 \right)$$

Let θ and ϕ be the mean angles of deviation in the electric and magnetic fields respectively,

$$\theta = S/l \quad \text{and} \quad \phi = S'/l'$$

$$\theta = \frac{eEl}{2mv^2} \qquad \phi = \frac{eBl'}{2mv}$$

or $$\theta = K_e \frac{e}{mv^2} \qquad ...(2)$$

$$\text{and } \phi = K_m \frac{e}{mv} \qquad ...(3)$$

where K_e and K_m are constants, depending on the strength and distribution of electric and magnetic fields.

Let $d\theta$ be the dispersion angle due to electric field and $d\phi$ the convergence angle due to the magnetic field. Differentiating equations (2) and (3), we get,

$$d\theta = -\ 2K_e \frac{e}{mv^3}\, dv = -\ 2\theta \frac{dv}{v}$$

or
$$\frac{d\theta}{\theta} = -\ 2 \frac{dv}{v} \qquad ...(4)$$

Similarly,

$$d\phi = -\ K_m \frac{e}{mv^2} dv$$

$$= -\phi \frac{dv}{v}$$

or
$$\frac{d\phi}{\phi} = -\frac{dv}{v} \qquad ...(5)$$

Comparing Eqs. (4) and (5), we get

$$\frac{d\theta}{\theta} = 2\frac{d\phi}{\phi} \qquad ...(6)$$

Let OR = a, and RL = b

The dispersion of the selected group at R is $ad\theta$. In the absence of magnetic field the dispersion produced in the beam for a distance a + b

$$= (a + b)\ d\theta$$

As this dispersion (or divergence) is annulled by the magnetic field acting at R so that the group comes to focus at L, distance b from R the condition of focusing is given by,

$$(a + b)\ d\theta = bd\phi$$

$$\frac{a+b}{b} = \frac{d\phi}{d\theta} \qquad ...(7)$$

$$\frac{a+b}{b} = \frac{\phi}{2\theta}, \text{ from eqn. (6)}$$

$$(a + b)\, 2\theta = b\phi$$

$$b\,(\phi - 2\theta) = 2a\theta \qquad ...(8)$$

This is the condition for focusing. If plate P is inclined at an angle α to the incident beam AB and RN be the perpendicular on the line joining O and plate P, from triangle RNO

$$RN = a \sin(\alpha + \theta) \qquad ...(9)$$

and from triangle RNL,

$$RN = b \sin RLN$$

$$RN = b \sin(\phi - \alpha - \theta) \qquad ...(10)$$

Equating eqns. (9) and (10),

$$a \sin(\alpha + \theta) = b \sin(\phi - \alpha - \theta)$$

for small angle,

$$a(\alpha + \theta) = b(\phi - \alpha - \theta) \qquad ...(11)$$

Comparison of eqns. (10) and (11) shows that two equations are identical if $\alpha = \theta$. *Thus the condition for focussing is that photographic plate must be placed at an angle θ with the direction of incident positive ray.*

Separation of Isotopes by Analysis of Traces

Substance of known atomic masses are mixed up with elements whose isotopes are to be determined and traces are obtained on photographic plate. The traces of the known atomic mass serve as reference, thus the traces of unknown isotopes are known by measuring the distance from the reference point.

Advantages of Aston's Mass Spectrograph Over Thomson Spectrograph

The modified technique of Aston have the following advantages over Thomson method:

1. The dispersion of the ionic beam according to different velocities of ions is greater than in Thomson mass spectrograph.
2. All the ions having the same e.m are focussed at one pint, instead of spreading them into a parabola thus securing greater intensity for photography. Hence a fine slit can be used to obtain sharp lines to be obtained on the photographic plate, thereby securing greater sensitivity. The accuracy was 1 in 1,000.

MAGNETIC FIELD OF A MOVING POINT CHARGE (LAPLACE RULE)

It was found experimentally that the magnetic induction resulting from a charge q moving with velocity $\vec{v}$ at a distance r away from the charge where r is a vector pointing from the charge to the point where the field is being found, is related by,

$$B \propto \frac{qv \sin\theta}{r^2}$$

where B is the magnitude of the induction and θ is the angle between $\vec{v}$ and $\vec{r}$ as shown in Fig. 3.16.

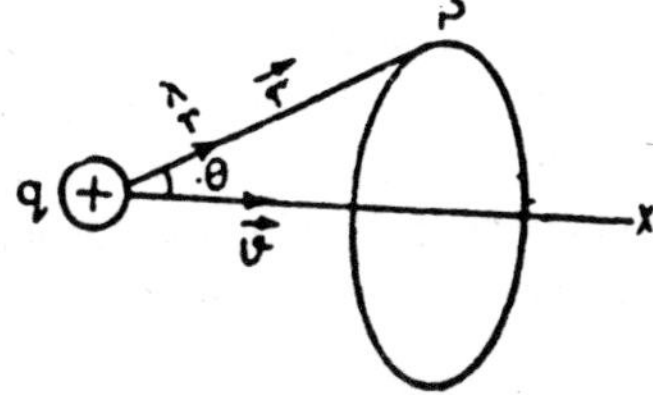

Fig. 3.16

The above equation can be written in terms of equality by inserting a constant $m_o/4\pi$. Therefore,

$$B = \frac{\mu_o}{4\pi} \frac{qv \sin\theta}{r^2}$$

or $$\vec{B} = \frac{\mu_o}{4\pi} q\left(\frac{\vec{v} \times \vec{r}}{r^2}\right) \quad ...(1)$$

or $$\vec{B} = \frac{\mu_o}{4\pi} q\left(\frac{\vec{v} \times \vec{r}}{r^3}\right) \quad ...(1a)$$

Where $\vec{r}$ is unit vector along r,

μ_o is called permeability of free space.

We have chosen the permeability constant $\frac{\mu_o}{4\pi}$ as for Coulomb's law, we introduced $\frac{1}{4\pi\varepsilon_o}$ and for the same reason *i.e.*, in rationalised MKS system the factors π should occur in problems having spherical, cylindrical or circular symmetry and should not occur when these symmetries do not exist. In fact, certain equations that are derived from equation (1) but are used more often than it is, will be simpler in form if we introduce $\frac{\mu_o}{4\pi}$. *We shall find the physical significance of the number just later.*

The direction of $\vec{B}$ is determined by the rules of vector product. From equation (1), $\vec{B}$ is perpendicular to $\vec{v}$ and $\vec{r}$, *i.e.*, in the direction normal to the plane containing $\vec{v}$ and $\vec{r}$ which is plane of the paper.

THE BIOT-SAVART LAW

The first significant relationship between a current and its magnetic field was discovered in 1820 by Oersted who found that a magnetised needle tends to set itself at right angle to a straight wire in which there is a steady current. In the same year, Biot and Savart formulated the equation for the field due to current in a long straight wire. Finally, they proposed a mathematical formula for the field due to a single current element rather than a finite length of wire, the equation can be used to calculate the field due to many different current configurations as shown in Fig. 3.17.

In fact, we are often interested not in the field of moving charge, but in that of an element of current, as a length dl of wire carrying a current I. We can easily set up an equivalence with equation (1). Suppose cross-section of wire is A and ρ is the charge density of charge in wire. Assume this charge moves with velocity v.

The charge crossing any cross-section per second = ρvA that is equal to current I

or $$\rho v A = I.$$

If we multiply by dl on both sides, we get,

$$\rho v A dl = I dl$$

$$qv = I dl$$

($\rho A dl$ is the total charge contained in wire of length dl.)

So the magnetic induction $d\vec{B}$ due to dl length of wire carrying current I is given by the following relation at a point P shown in Fig. 3.17. After substituting qv = Idl in equation (1), we get,

$$d\vec{B} = \frac{\mu_o}{4\pi} I \left(\frac{\vec{dl} \times \hat{r}}{r^2} \right) \qquad ...(2)$$

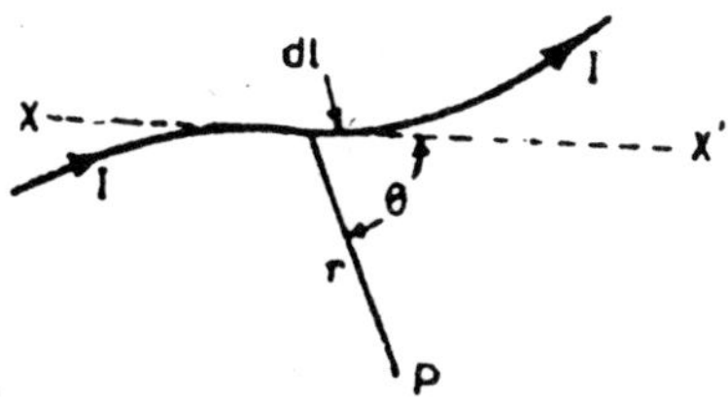

Fig. 3.17

or,
$$d\vec{B} = \frac{\mu_0}{4\pi} I\left(\frac{\vec{dl} \times \vec{r}}{r^3}\right) \quad ...(2a)$$

The law described by equations (1) and (2) is called Biot-Savart law.

Principle of Superposition

The resultant field at P is found by integrating equation (4.2) as the principle of superposition also holds here as in the case of electric field,

$$\vec{B} = \oint d\vec{B}$$

or
$$\vec{B} = \frac{\mu_0}{4\pi}\oint \frac{I}{r^3}\left(\vec{dl} \times \vec{r}\right) \quad ...(4.3)$$

where $\oint$ stands for close path.

The direction of $d\vec{B}$ is normal to the line from the current element Idl to the observation point P. (By the rule discussed in last chapter).

Right Hand Rule

Grasp the wire with right hand, the thumb pointing in the direction of current, the fingers will curl around the wire in the direction of $\vec{B}$.

THE FORCE BETWEEN TWO FINITE ELEMENTS OF CURRENT

Consider two current elements $I_1 dl_1$ and $I_2 dl_2$ flowing in different conductors as shown in Fig. 3.18. The magnetic induction due to I_2 current at a distance r on dl_1 is,

$$d\vec{B} = \frac{\mu_0}{4\pi} I_2\left(\frac{\vec{dl}_2 \times \vec{r}}{r^3}\right)$$

Now the force on conductor carrying current I_1, say dF, due to current I_2 can be calculated by using the relation $(\vec{F} = I\vec{l} \times \vec{B})$.

Thus the force between two elements of current $I_1 dl_1$ and $I_2 dl_2$ is given by,

$$d\vec{F} = \frac{\mu_0 I_1 I_2}{4\pi}\left(\frac{d\vec{l}_1 \times (d\vec{l}_2 \times \vec{r})}{r^3}\right) \quad ...(1)$$

This law for current element is equivalent to Coulomb's law of force between static charges.

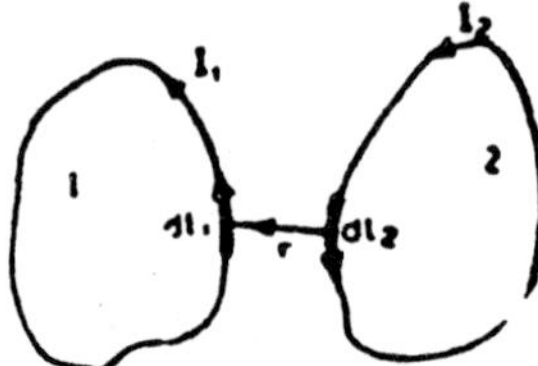

Fig. 3.18

Biot-Savart Law and Coulomb's Law of Electrostatics

Eqs. 1 and 2 reveal that Biot-Savart law is similar to Coulomb's law in electrostatics. The field or force varies inversely as the square of the distance from the current element. It differs from Coulomb's law, since the direction of field is normal to the line from the current element (Idl) to the point of observation, instead of being directed along this line, s in Coulomb's law.

APPLICATIONS OF BIOT-SAVART LAW

we shall calculate the B in some important cases as given below:

1. Magnetic Field due to Steady Current in a Long Straight Wire

As an example of use of Biot-Savart law, let us calculate the magnetic field vector $\vec{B}$ at a point P a distance R from a long straight wire in which there is a current I as shown in Fig. 3.19. The magnetic field $d\vec{B}$ due to current element $Id\vec{l}$ (*i.e.*, dl length of the wire carrying current I) of the wire at point P is given by Biot-Savart law *i.e.*,

$$d\vec{B} = \frac{\mu_0}{4\pi} I\left(\frac{d\vec{l} \times \vec{r}}{r^2}\right) = \frac{\mu_0}{4\pi}\frac{Idl}{r^2}\left(\hat{j} \times \hat{r}\right)$$

where $\hat{j}$ is unit vector along y-axis and
$\hat{r}$ is unit vector along r.

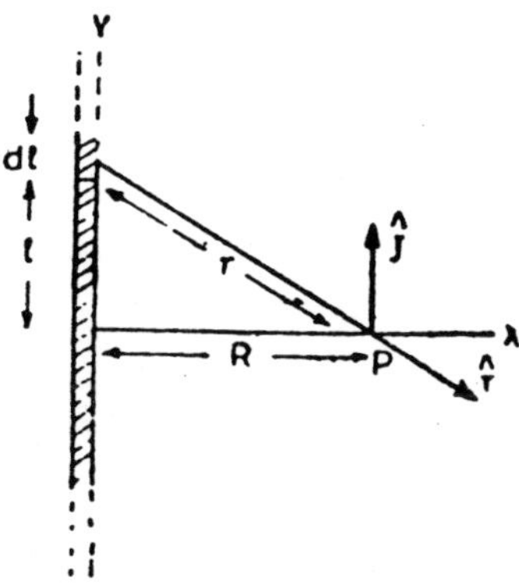

Fig. 3.19

or $$dB = \frac{\mu_o}{4\pi} \frac{Idl \sin \theta}{r^2}$$

The direction of $d\vec{B}$ at point P for all elements are the same, that is, into the plane of the figure at right angle to the page. The magnitude of $d\vec{B}$ due to whole wire is given by,

$$B = \int dB = \frac{\mu_o I}{4\pi} \int_{l=-\infty}^{l=+\infty} \frac{\sin\theta \, dl}{r^2}$$

From Fig 3.19,

$$r = (l^2 + R^2)^{1/2}$$

and , $$\sin \theta = \sin (\pi - \theta) = R/r$$

$$= \frac{R}{(l^2 + R^2)^{\frac{1}{2}}}$$

Putting these values, the expression for B becomes,

$$B = \frac{\mu_o I}{4\pi} \int_{-\infty}^{+\infty} \frac{Rdl}{(l^2 + R^2)^{\frac{3}{2}}} \qquad ...(1)$$

To evaluate the integral on R.H.S., substitute,

$$l = R \tan \phi, \qquad dl = R \sec^2 \phi \, d\phi$$

therefore, $$B = \frac{\mu_0 I}{4\pi R} \int_{-\pi/2}^{+\pi/2} \cos\phi \, d\phi$$

$$= \frac{\mu_0 I}{2\pi R} \qquad ...(2)$$

The direction of $\vec{B}$ is specified by $(\hat{j} \times \hat{r}) = \hat{\theta}_0$ and is perpendicular to the plane containing $\hat{j}$ and $\hat{r}$. Hence,

$$\vec{B} = \frac{\mu_0 I}{2\pi R} \hat{\theta}_0 \qquad ...(2a)$$

Also the direction of $\vec{B}$ by R.H. rule i.e, grasp the wire with right hand, the thumb pointing in the direction of current, the finger will curl around the wire in the direction of B, hence magnetic field lines are circular closed curves around the wire.

2. Magnetic Field Near a Straight Current Filament of Finite Length

Let us calculate the magnetic field due to a wire of length $2l$ at a point P as shown in Fig. 3.20. The geometry and notations to be used are shown in Fig. 3.20. The wire is placed along x-axis and origin of the coordinate system is at the middle. Further, let the point P be at a distance R from the x-axis.

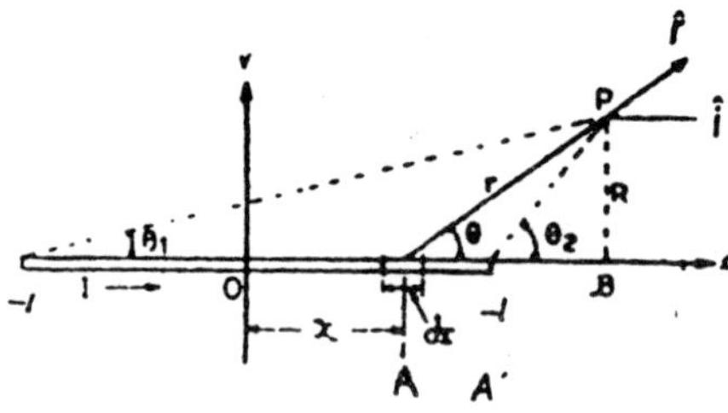

Fig. 3.20

Consider the contribution to the magnetic field by the current element Idx at P as shown in Fig. 3.20. According to equation (2(a)) the magnetic field $d\vec{B}$ is given by,

$$d\vec{B} = \frac{\mu_0}{4\pi} I \left(\frac{d\vec{x} \times \hat{r}}{r^2} \right)$$

Form figure $\quad r^2 = AB^2 + R^2,$

and $\quad AB = AA' + A'B$

where $AA' = l - x$ and

$A'B = R \cot \theta_2$

Hence $AB = l - x + R \cot \theta_2$

So $r2 = (l - x + R \cot \theta_2)^2 + R^2$, therefore,

$$d\vec{B} = \frac{\mu_o}{4\pi} \frac{Idx}{(l - x + R \cot \theta_2)^2 + R^2} (\hat{i} \times \hat{r})$$

where, $d\vec{x} = \hat{i}dx$, and

$\hat{i}$ = unit vector along x-axis

From the figure, it is seen that in triangle APB,

$$l - x + R \cot \theta_2 = R \cot \theta$$

then $dx = R \operatorname{cosec}^2 \theta \, d\, \theta$

and $\operatorname{cosec} \theta = \frac{r}{R} \frac{\sqrt{(l - x + R \cot \theta_2)^2 + R^2}}{R}$

and $dx = \frac{(l - x + R \cot \theta_2)^2 + R^2}{R} d\theta$

Substituting dx in the expression $d\vec{B}$, we get,

$$d\vec{B} = \frac{\mu_o}{4\pi} \frac{I}{R} d\theta \left(\hat{i} \times \hat{r}\right)$$

But the magnitude of the product $\hat{i} \times \hat{r}$ is $\sin \theta$ and direction is perpendicular to plane containing $\hat{i}$ and $\hat{r}$,

i.e., $\hat{i} \times \hat{r} = \hat{\theta}_o \sin \theta d$

where $\hat{\theta}_o$ is unit vector perpendicular to the plane containing $\hat{i}$ and $\hat{r}$, hence,

$$d\vec{B} = \frac{\mu_o}{4\pi} \frac{I}{R} \sin \theta \, d\theta \, \hat{\theta}_o \qquad ...(2b)$$

Magnitude $dB = \frac{\mu_o I}{4\pi R} \sin \theta \, d\theta$

The total field is obtained by integrating between the limit sq1 and θ_2 due to length 2l of the wire,

$$B = \int_{\theta_1}^{\theta_2} dB \quad \int_{\theta_1}^{\theta_2} \frac{\mu_o I}{4\pi R} \sin \theta \, d\theta$$

$$= \frac{\mu_o I}{4\pi R} \left| -\cos\theta \right|_{\theta_1}^{\theta_2}$$

$$B = \frac{\mu_o I}{4\pi R}(\cos\theta_1 - \cos\theta_2)$$

and directed along $\hat{\theta}_o$, hence,

$$\vec{B} = \frac{\mu_o I}{4\pi R}(\cos\theta_1 - \cos\theta_2)\,\hat{\theta}_o \qquad \text{...(2c)}$$

Consequently, the magnetic field is circular and by symmetry has its centre on the axis. For the special case when the length becomes very long, there by approaching the infinite case, *i.e.*,

$$l \to \infty,\ \theta_1 \to 0, \text{ and } \theta_2 \to \pi$$

then,

$$\vec{B} = \frac{\mu_o I}{2\pi R}\hat{\theta}_o \qquad \text{...(2d)}$$

which is same as given in equation (2(d))

3. Interaction Between Two Parallel Long Currents

Fig. 3.21 shows tow long parallel wires separated by distance d and carrying currents I_a and I_b. It is an experimental fact, noted by Ampere only one week after Oersted's magnetic phenomena of electric current. Wire a will produce a field of induction B_a at all nearby points. The magnitude of B_a, due to current I_a at a distance d *i.e.*, on wire b is,

$$B_a = \frac{\mu_o I_a}{2\pi d}$$

According to the R.H. rule the direction of B_a is downward as shown in Fig. 3.21. Wire b, which carries a current I_b is now really placed in such field. Consider the *l* length of the wire b it will experience a force $(I\vec{l} \times \vec{B})$ whose magnitude is,

$$F_b = I_b l B_a = \frac{\mu_o l I_a I_b}{2\pi d}$$

(After substituting Ba given above) ...(3)

Direction of F_b is determined according to vector rule F_b lies in the plane of the wires and points to the left.

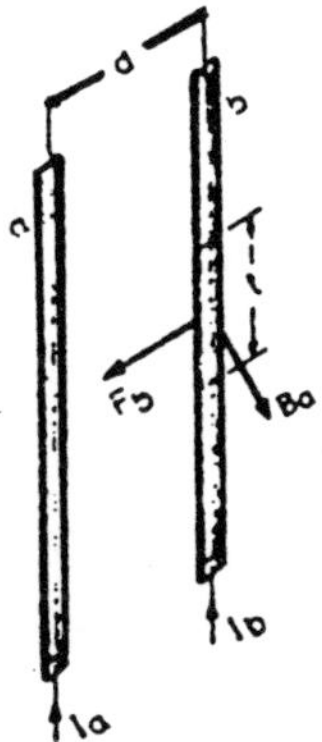

Fig. 3.21

One can start with wire b, compute the field of induction which it produces at the site of wore a, and then compute the force on wire a, *for parallel currents, F_a points to the right. The forces that the two wires exert on each other are equal and opposite, hence according to Newton's* law of action and reaction, *they will attract. For antiparallel currents, the two wires repel each other.* We can compare the above discussion with that of two point charges. There we explained that two charges act on each other through the intermediary of the electric field. Here two conductors in same manner by; magnetic field. We think as,

$$\text{current} \rightleftarrows \text{field}$$

and not as in the action at a distance point of view in terms of,

$$\text{current} \rightleftarrows \text{current.}$$

Unit of Current in Rationalised MKS

The force of attraction between two very long parallel wires is used to define the ampere practically. The ampere is that current which when flowing through two long parallel straight conductors placed in vacuum at a distance of one metre from each other produces a force of attraction of 2×4^{-7} Newtons per metre length between the wires, *i.e.*, when in equation (4),

$$l = 1 \text{ metre}, I_a = I_b = 1 \text{ amp}, d = 1 \text{ metre}$$

$$F = 2 \times 10^{-7} \text{ Newtons}$$

Hence $$\mu_o = 4\pi \times 10^{-7} \text{ Newtons/amp}^2.$$

$$= 4\pi \times 10^{-7} \frac{\text{Newtons. metre}}{\text{Ampere}^2\text{. metre}}$$

$$= 4\pi \times 10^{-7} \frac{\text{Weber}}{\text{Amp. metre}}$$

Since, $$1 \text{ Weber} = \frac{1 \text{ Newton. metre}}{1 \text{ Ampere}}$$

'Basically' the ampere is defined by equation (5). Thus if we take two current elements each of unit length, and at unit distance apart, so oriented that the factor,

$$\frac{d\vec{l}_1 \times \left(d\vec{l}_2 \times \vec{r}\right)}{r^3} = 1$$

Then the currents are defined to an ampere, *i.e.*,

$$I_1 = I_2 \, 1 \text{ amp}$$

if the resulting force is 10^{-7} Newtons. From this it follows by equation (5),

$$\mu_o = 4\pi \times 10^{-7} \text{ Newton/ampere}^2$$

$$\mu_o = 4\pi \times 10^{-7} \text{ Weber/amp.metre}$$

Be Careful to Note The Following Facts

The value assigned to μ_o is seen to be the result of definition rather than experiment. This, on the other hand is not the case with ε_o, which we encountered in Coulomb's law of electrostatics. The difference is that we define th unit of current, the ampere, from the Biot-Savart law, whereas, we define the coulomb, the unit of charge, as the charge flowing per second in a current of 1 ampere. Therefore, we are not free to choose the constant *i.e.*, ε_o in Coulomb's law at our own choice but must determine it by experiment.

4. Magnetic Field Along The Axis of Circular Coil

Consider a circular coil of radius R and carrying a current I.

The vector $d\vec{l}$ for a current element at the top of the coil points perpendicular towards the reader. The $d\vec{l}$ and $\vec{r}$ are perpendicular to each other and plane formed by $d\vec{l}$ and $\vec{r}$ is normal to the page as shown in Fig. 3.22.

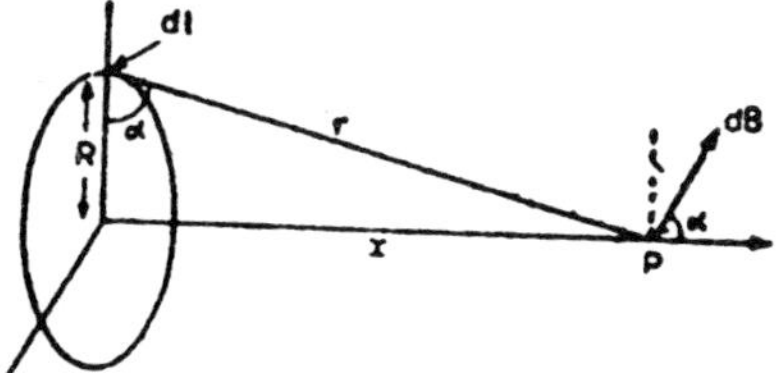

Fig. 3.22

The $d\bar{B}$ for this element is at right angle to this plain, this lies in the plane of the figure and perpendicular to r. The magnitude of dB from equation (6) is given by

$$dB = \frac{\mu_0}{4\pi} \frac{Idl}{r^2} \quad ...(6)$$

Let us resolve $d\bar{B}$ into two components, one, $d\vec{B}\|$, along the axis of loop and another, $dB\perp$,at right angles to x-axis. As the coil is symmetrical about the x-axis the contribution $d\bar{B}$ due to element of coil on opposite side the $dB\perp$ will be equal in magnitude but opposite in direction and therefore, cancel out, we have only $dB\,\|$ component. The resultant B for the complete loop is given by,

$$B = \int dB$$

$$= \int \frac{\mu_0 I}{4\pi R} dl \cos\alpha$$

Since $dB\| = dB\cos\alpha$ and dB is given in Eq. (7) and from figure.

$$r = \sqrt{R^2 + x^2}$$

$$\cos\alpha = R/r = R/\sqrt{R^2 + x^2}$$

therefore,

$$B = \int \frac{\mu_0 I}{4\pi} \frac{Rdl}{(R^2 + x^2)^{3/2}}$$

$$= \frac{\mu_0 IR}{4\pi(R^2 + x^2)^{3/2}} \int dl$$

$\int dl$ is the circumference of the loop $2\pi R$.

$$B = \frac{\mu_0 IR^2}{2(R^2 + x^2)^{3/2}} \quad ...(8)$$

If coil having N turns, then

$$B = \frac{\mu_0 NIR^2}{2(R^2 + x^2)^{3/2}} \quad ...(8a)$$

This shows the B decreases as the distance x increases. At the centre $x = 0$

$$B_{centre} = \frac{\mu_0 IR^2}{2R^3} = \frac{\mu_0 I}{2R} \quad ...(9)$$

If $x >> R$ *i.e.*, point P lies far from coil in Eq. (8), R in denominator can be neglected. Hence,

$$B = \frac{\mu_0 IR^2}{2x^3} \quad ...(10)$$

If coil having N turns, then,

$$B = \frac{\mu_0 NIR^2}{2x^3} \quad ...(11)$$

Recalling the area of the coil is πR^2,

$$B = \frac{\mu_0 (NIA)}{2\pi x^3}, \ (A = \pi R^2)$$

and NIA according to definition given in the last magnetic dipole moment of the current coil represented by m

$$B = \frac{\mu_0 m}{2\pi x^3} \quad ...(12)$$

Variation of The Along The Axis of Coil and Principle of Helmholtz Galvanometer

Differentiating Eqn. (8) with respect to x, we get,

$$\frac{dB}{dx} = \frac{\mu_0 IR^2}{2} (-3/2).\ 2x(R^2 + x^2)^{-5/2}$$

$$\frac{d^2B}{dx^2} = -\frac{3\mu_0 IR^2}{2} [(R_2 + x^2)^{-5/2} + x(-5/2).\ 2x\ (R^2 + x^2)^{-7/2}]$$

or, $$\frac{d^2B}{dx^2} = -\frac{3\mu_0 IR^2}{2} (R^2 + x^2)^{-7/2} [R^2 + x^2 - 5x^2]$$

$$\frac{d^2B}{dx^2} \text{ is zero when } R^2 + x^2 - 5x^2 = 0$$

or $$x = \pm \frac{R}{2} \qquad ...(13)$$

and at those points $\frac{d\vec{B}}{dx}$ is constant. Hence, at a distance of $\frac{R}{2}$ on either side of the coil, the rate of change of the field becomes constant. This is the principle of Helmholtz galvanometer. The variation is shown in Fig. 3.23.

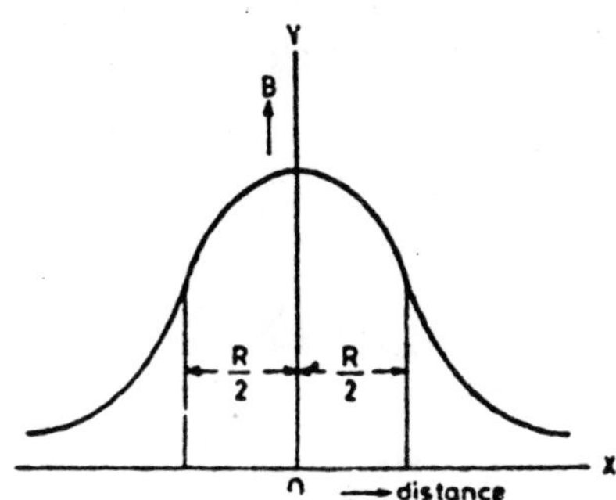

Fig. 3.23

Helmholtz Galvanometer

It is improved form of the tangent galvanometer. It consists of two equal coils placed coaxially at a distance equal to the radius of the either coil as shown in Fig. 3.24(a).

The two coils are of the same radius and are connected in series so that the current through them is in the same direction. The field at any point on the axis of the coil is equal to the sum of the magnetic field due to individual coil.

Since the rate of change of field is constant at a distance R/2 from each coil as we discussed in previous article (Eq. 13). Therefore, *the field at a point mid-way between the coils is most uniform. As we pass away from this point, the increase in the field due to this coil is compensated by an equal decrease in the field due to the other coil.*

Thus the field is quite uniform over a considerable distance and magnetic needle rotates in this uniform magnetic field. And it is not necessary to use a short magnetic needle as in the case of tangent galvanometer. So that we can use a large magnetic needle, which provides a better accuracy in measuring the current.

Fig. 3.24(a)

The magnetic field at the middle point between the two coils is obtained from equation (8). Since there are two coils, replacing N by 2`J and putting x = R/2. Thus

$$\vec{B} = \frac{\mu_0 2NIR^2}{2\left(R^2 + \frac{R^2}{4}\right)^{3/2}}$$

$$= \frac{8}{\sqrt{125}}\left(\frac{\mu_0 NI}{R}\right) \qquad \text{...(13a)}$$

If H be the horizontal component of the earth's magnetic field and needle shows a deflection θ then by tangent law,

$$B = H \tan \theta$$

therefore, $$\frac{8}{\sqrt{125}}\left(\frac{\mu_0 NI}{R}\right) = H \tan \theta$$

or $$I = \frac{\sqrt{125}}{8\mu_0 N} RH \tan \theta \qquad \text{...(13b)}$$

If Fig. 3.24(b) the curve representing the variation of magnetic fields of two coils with distance.

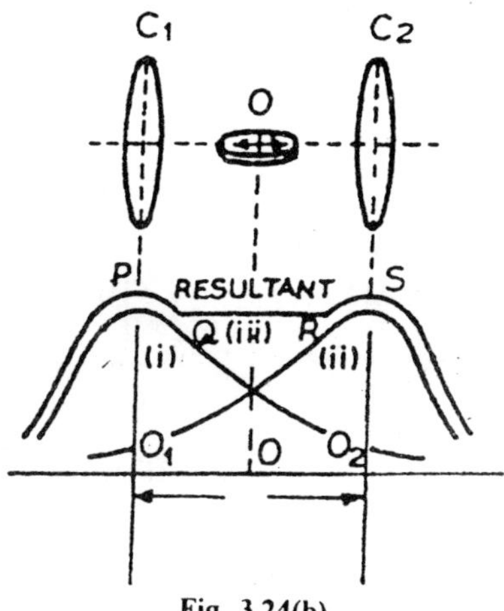

Fig. 3.24(b)

The curve PQRS give the resultant field and is obtained by adding the ordinates of two curves at every point. The straight portion QR of this curve indicates a uniform field near the point O, *i.e.*, middle point of the two coils.

5. Field on Axis of Solenoid

The result of Eqn. (14) may be extended to find vector $\vec{B}$ at a point P on the axis of a solenoid. A coil of wire wound in the form of cylinder is called solenoid, Fig. 3.25(a). Assuming the uniform winding, so that the number of terms per unit length of the solenoid are constant, say, n.

Let us first calculate the field due to a small segment of length dx of the solenoid which subtends angle θ and θ + dθ with the axis at the point P. From Fig. 3.25(b) we have,

$$\frac{rd\theta}{dx} = \sin\theta$$

or

$$dx = \frac{rd\theta}{\sin\theta}$$

The number of turns on the length dx $= \dfrac{nrd\theta}{\sin\theta}$

and the total current in this element dx $= \dfrac{Inrd\theta}{\sin\theta}$

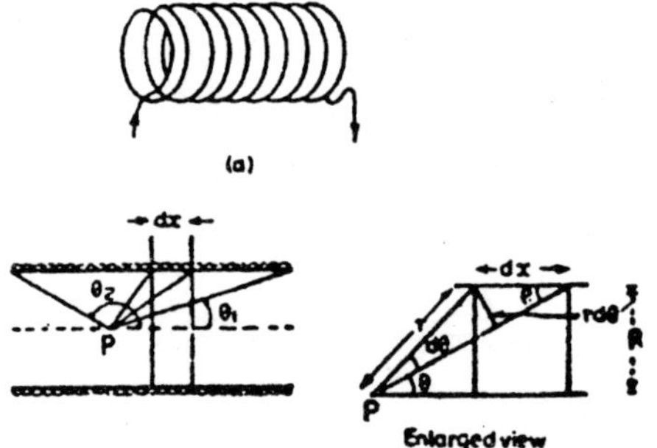

Fig. 3.25(a). Fig. 3.25(b).

where I is the current flowing in the solenoid putting this n Eqn. (14) in place of I we get the contribution of the current rings in a length dx of the solenoid. The direction of the field is along the axis of coil that is x-axis hence dB due to dx element is given by,

$$dB = \frac{\mu_0 R^2}{2r^3} \frac{Inrd\theta}{\sin\theta}, \text{ since } \sin\theta = R/r$$

we get

$$= \frac{\mu_0 In \sin\theta \, d\theta}{2}$$

The field due to whole solenoid is

$$B = \frac{\mu_0 In}{2} \int_{\theta_1}^{\theta_2} \sin\theta \, d\theta$$

$$= \frac{\mu_0 In}{2} (\cos\theta_1 - \cos\theta_2) \qquad ...(15)$$

Note that if the fingers of the right hand are curled around the solenoid in the direction of circulations of the current, the thumb will point in the direction of B.

If the solenoid is very long and P is some where near the middle, $\theta_1 = 0$, and $\theta_2 = \pi$ then,

$$B = \mu_0 In \qquad ...(16)$$

If the point P lies at one end of the solenoid, $\theta_1 = 0$, and $\theta_2 = \pi/2$ and thus,

$$B = \mu_0 n I/2 \qquad ...(17)$$

This shows that $\vec{B}$ *at the end of a long solenoid is just one half of that at the centre,* the variation is shown in Fig. 3.25(c) and it is clear that B at the centre of the solenoid remains nearly constant until we reach one of the ends.

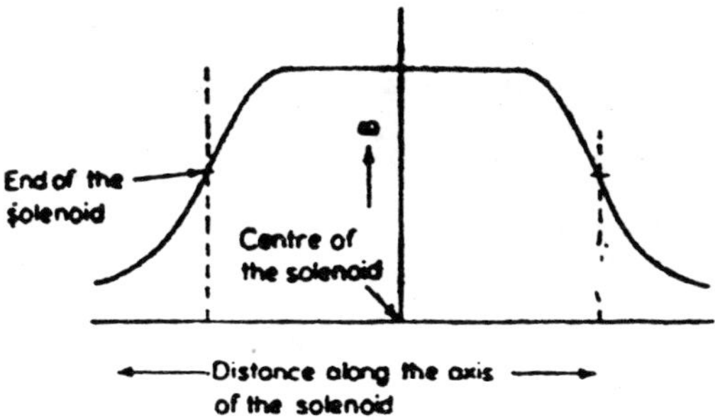

Fig. 3.25(c)

SOLVED EXAMPLES

Example 1:

If a stream of electrons, each of mass m and charge e and velocity 3×10^6 *metre/sec. is deflected by 2 mm in traversing a distance of 10cm through an electrostatics field of 18 V/cm. perpendicular to their path, find e/m in couls/kg.*

Solution:

$$E = 18\,\frac{\text{Volts}}{\text{cm}} = \frac{18}{10^{-2}}\,\frac{\text{Volts}}{\text{metre}}$$

$$eE = ma$$

$$y = \frac{1}{2}at^2 = \frac{1}{2}\left(\frac{eE}{m}\right)\left(\frac{1}{v}\right)2$$

$$2\times10^{-3} = \frac{1}{2}\left(\frac{18}{10^{-2}}\cdot\frac{e}{m}\right)\left(\frac{10\times10^{-2}}{3\times10^{6}}\right)^2$$

or $$\frac{e}{m} = \frac{2\times10^{-3}\times2\times10^{-2}\times9\times10^{12}}{18\times100\times10^{-4}}$$

$$= 2\times10^{11}\ \text{couls. kg.}$$

Example 2:

For a proton initially travelling with velocity vx = 5 × 10⁸ cm/sec. Calculate the following:

(1) *Force experienced in a magnetic field* $\vec{B} = 2000\ \hat{j} + 4000\ \hat{k}$ *gauss*

(2) *Acceleration experienced in an electric field E,*

$$\vec{E} = 200\hat{i} + 100\hat{j} \text{ volts/cm}$$

(3) *Transverse deflection in travelling a length* $L_x = 10$ *cm in an electric field* $E_y = 200$ *volts/cm.*

Solution:

(1) Force in magnetic field

$$\vec{F} = q(\vec{v} \times \vec{B})$$

$$\vec{v}_x = 5 \times 10^8 \text{ cm/sec} = 5 \times 10^6 \text{ metre/sec}$$

$$\vec{B} = 2000\ \hat{j} + 4000\ \hat{k}$$

$$\vec{B} = 0.2\ \hat{j} + 0.4\ \hat{k} \text{ weber/m}^2$$

Since, $\dfrac{1 \text{ Weber}}{\text{metre}^2} = 10^4$ gauss

hence,

$$\vec{F} = 1.6 \times 10^{-19} [(5 \times 10^6\ \hat{i}) \times (0.2\ \hat{j} + 0.4\ \hat{k})]$$

$$= 1.6 \times 10^{-19} \times 5 \times 10^6 \times 2\ [0.1\hat{k} - 0.2\ \hat{j}]$$

$$= 1.6 \times 10^{-13}\ [\ \hat{k} - 2\hat{j}] \text{ Newtons}$$

Magnitude of the force F = 1.6×10^{-13} Ö5 = 3.57×10^{-13} Nt.

Its direction cosines

$$\left((0, -\frac{2}{\sqrt{5}}, \frac{1}{\sqrt{5}}\right)$$

(2) In electric field,

$$ma = qE$$

$$a = \frac{qE}{m} = \frac{1.6 \times 10^{-19}}{1.67 \times 10^{-27}} \left(\frac{200\hat{i} + 100\hat{j}}{10^{-2}}\right)$$

$$= \frac{1.6}{1.67} 10^{12}\ (2\hat{i} + \hat{j})$$

$$= \frac{1.6}{1.67} 10^{12} \sqrt{5}$$

$$= 2.14 \times 1012 \text{ met/sec}^2$$

(3) Transverse deflection,

$$Y = \frac{1}{2} at^2$$

$$= \frac{1}{2}\left(\frac{gE_y}{m}\right)\left(\frac{l}{v_x}\right)^2,$$

$$\text{Since } E_y = 200 \frac{\text{volt}}{\text{cm}} = \frac{200}{10^{-2}} \frac{\text{volt}}{\text{metre}}$$

$$Y = \frac{1}{2} \times \frac{1.6 \times 10^{-19} \times 200}{1.67 \times 10^{-27} \times 10^{-2}} \times \left(\frac{10 \times 10^{-2}}{5 \times 10^{-6}}\right)^2 \text{ metre}$$

$$= 0.0383 \times 10^{-2} \text{ metre}$$

$$= 0.0383 \text{ cm}$$

Example 3:

An electron of energy 20 eV moving in a direction perpendicular to a magnetic field B = 10² weber/m² describes a circle or radius R. Calculate the value of R.

Solution:

Kinetic energy $\frac{1}{2} mv^2 = 20 \text{ eV}$

$$= 20 \times 1.6 \times 10^{-19} \text{ joules}$$

hence, $v = \left(\frac{2 \times 20 \times 1.6 \times 10^{-19}}{m}\right)^{1/2} \text{ m/sec}$

The radius of the circular path is given by equation (1),

$$R = \frac{mv}{qB}$$

$$= \frac{m \times \left(64 \times 10^{-19}\right)^{1/2}}{\sqrt{m \times q \times B}}$$

$$= \frac{\left(9 \times 10^{-31} \times 64 \times 10^{-19}\right)^{1/2}}{1.6 \times 10^{-19} \times 10^2}$$

$$= 15.0 \times 10^{-8} \text{ metre}$$

Example 4(a):

An electron of velocity $\vec{v} = (2\hat{i} + 3\hat{j})\ 10^8$ cm/sec. enters a region of uniform magnetic field, $B = 500\ \hat{i}$ gauss, show that its path becomes helical.

(1) In what direction does the axis of the helix lie ?

(2) Calculate the radius of helix

(3) Calculate the number of rotations as the electron advances 10 cm along the axis of the helix.

Solution:

$\vec{v} = (2\hat{i} + 3\hat{j})\ 10^8$ cm/sec. and

$\vec{B} = 500\hat{i}$ gauss

The component of velocity $\vec{v}_1 = 2\hat{i} \times 10^8$ cm/sec. is along $\vec{B}$ and electron moves along x-axis and due to the component $\vec{v}_2 = 2\hat{j} \times 10^8$ cm/sec. which is perpendicular path.

(1) The axis of helix lines along B *i.e.*, x-axis.

(2) Radius of helix R by equation (1)

$$R = \frac{mv_2}{qB} + \frac{9\times10^{-31}\times3\times10^{6}}{1.6\times10^{-19}\times500\times10^{-4}}$$

(1 Weber/metre^2 = 10^4 gauss)

$$R = 3.37 \times 10\text{–}4 \text{ metre}$$

$$= 0.0337 \times 10\text{–}2 \text{ metre}$$

$$= 0.0337 \text{ cm}$$

(2) Distance travelled in one rotation = path of the helix by equation (2),

$$\frac{2\pi m\, v_1}{qB} = \frac{2\pi\times9\times10^{-31}\times2\times10^{6}}{1.6\times10^{-19}\times500\times10^{-4}}$$

Number of rotations made by the electron as it advances through a distance of 10 cm

$$= \frac{10\times10^{-2}}{\text{pitch of helix}}$$

$$= \frac{10\times10^{-2}\times1.6\times10^{-19}\times500\times10^{-4}}{2\pi\times9\times10^{-31}\times2\times10^{6}}$$

$$= 70.74$$

$$= 71.$$

Example 4(b):

A long straight wire carries a current of 20 amperes. An electron is travelling at 10^7 m/sec, is 2.0 cm from the wire, what force acts on the electron if its motion is directed (1) towards the wire, (2) parallel to the wire and, (3) at right angles to the direction given in (1) and (2)

Solution:

The magnitude of B produced by wire carrying currents is,

$$B = \frac{\mu_0 I}{2\pi r} \frac{4\pi \times 10^{-7} \times 20}{2\pi \times 0.02/100}$$

$$= 2 \times 10^{-4} \text{ weber/metre}^2$$

Force on the electron $\vec{F} = q(\vec{v} \times \vec{B})$

(1) When electron moves towards the wire, v is perpendicular to the B due to wire so,

$$F = qvB = 1.6 \times 10^{-19} \times 10^7 \times 2 \times 10^{-4}$$

$$= 3.2 \times 10^{-16} \text{ Newtons.}$$

F is parallel to the current.

(2) When electron moves parallel to the wire, its velocity v will again be perpendicular to the magnetic field B. So the magnitude will be $= 3.2 \times 10^{-16}$ Newtons.

(3) When the electron is moving at right angle to the direction given above then v and B will be parallel or anti-parallel.

Hence, $\vec{F} = q(\vec{v} \times \vec{B})$ will be zero

or $F = 0$

Example 4(c):

A solenoid is 1 metre long an d3 cm in mean diameter. It has 5 layers of windings of 850 turns each and carries a current of 5.00 amp. Calculate B at its centre.

Solution:

$$B = \mu_o I_o n$$

$$= 4\pi \times 10^{-7} \times 5 \times 5 \times 850$$

$$= 2.7 \times 10^{-2} \text{ weber/metre}^2$$

Example 5:

The electron circulates around the nucleus in a path of radius 5.1 × 10⁻¹¹ m at a frequency of 6.8 × 10¹⁵ revolutions/second. Calculate B at the centre and the magnetic dipole moment.

Solution:

$$\text{Current I} = \text{Charge flow/sec} = \text{ev}$$
$$= 1.6 \times 10^{-19} \times 6.8 \times 10^{15}$$
$$= 1.1 \times 10^{-3} \text{ amp.}$$

B at the centre of the orbit is given by putting x = 0 in Eq. (1).

$$\text{i.e., } B = \frac{\mu_0 IR^2}{2(R^2+x^2)^{3/2}} = \frac{\mu_0 I}{2R} = \frac{4\times10^{-7}\times1.1\times10^{-3}}{2\times5.1\times10^{-11}}$$
$$= 14 \text{ weber/metre}^2$$

Magnetic dipole moment from Eq. (2) of Chapter 9,

$$m\ NIA = 1 \times 1.1 \times 10^{-3} \times \pi \times (5.1 \times 10^{-11})^2$$
$$= 9 \times 10^{-24} \text{ amp.metre}^2.$$

Example 6(a):

Two long parallel wires separated by 2 cm in air carry current of 100 amp. Find the force on one metre length of a conductor.

Solution:

Using relation (1), *i.e.*,

$$F = \frac{\mu_0 l I_a I_b}{2\pi d}$$

As given $l = 1$ metre

$I_a = I_b$ 100 amp.

$d = 2 \text{ cm} = 2 \times 10^{-2}$ metre

and putting $\mu_o = 4\pi \times 10^{-7}$ weber/amp. metre

$$F = 4\pi\times10^{-7}\ \frac{\text{weber}}{\text{amp.m}}\times\frac{1}{2\pi}\times\frac{1\text{m}\times100\times100\text{amp}^2}{2\times10^{-2}\text{ metre}}$$
$$= 0.1\ \frac{\text{weber.amp}}{\text{metre}}$$
$$= 0.1 \text{ Newton}$$

Example 6(b):

Find an expression for the magnetic induction at the centre of a circular current loop.

Solution:

Magnetic field at the centre of a circular current loop.

Consider a circular current loop of radius R carrying current I. Let us first calculate the magnetic from $\vec{B}$ due to small element $d\vec{l}$.

$$d\vec{l} = R d\theta\, \hat{\theta}_o$$

where is the a unit vector in the θ direction. Note that since $d\vec{l}$ is small means dθ is small. The unit vector.

$\hat{r}$ and $\hat{\theta}_o$ are perpendicular to each other at all points along the path. The, from equation (2), the field due to this element at the centre,

$$d\vec{B} = \frac{\mu_o}{4\pi}\frac{I(d\vec{l} \times \hat{r})}{R^2}$$

$$= \frac{\mu_o I}{4\pi}\left(\frac{Rd\theta\, \hat{\theta}_o \times \hat{r})}{R^2}\right)$$

$$= \frac{\mu_o}{4\pi} I \frac{Rd\theta}{R^2}(\hat{\theta}_o \times \hat{r})$$

Hence, total $\vec{B}$ due to coil at centre O,

$$B = \int_0^{2\pi} d\vec{B} = (\hat{\theta}_o \times \hat{r})\frac{\mu_o}{4\pi}\int_0^{2\pi}\frac{Id\theta}{R}$$

$$= (\hat{\theta}_o \times \hat{r})\frac{\mu_o}{4\pi}\frac{2\pi I}{R}$$

$$= \frac{\mu_o I}{2R}\hat{k}$$

$\hat{\theta}_o \times \hat{r} = \hat{k}$ is a unit vector along z-axis because B is directed along the direction perpendicular to the plane containing $\hat{\theta}_o$ and $\hat{r}$ *i.e.*, xy plane here.

Hence, B is directed out of the plane of the paper, *i.e.*, z-axis.

Example 6(c):

A current of 10 amp. flows in each of the two conducting wires parallel to each other. The separation between the wires is 2 cm. Find the force per unit length of one of the wires. Will it be a force of attraction or repulsion ?

Solution:

$$F = \frac{\mu_0 I_1 I_2}{2\pi r} \text{ Newtons/metre}$$

Here $I_1 = I_2 = 10$ amp. $R = 2$ cm $= 0.02$ m,

therefore, $$F = \frac{(4\pi \times 10^7 \text{ Newtons/amp}^2)(10 \text{ amp})(10 \text{amp})}{2\pi \times (0.02 \text{m})}$$

$$= 10^{-3} \text{ Newtons/m.}$$

Example 7:

Two similar coils of wire having a radius of 7 cm. and 60 turns have a common axis and are 18 cm apart. Find the strength of the magnetic field at a point midway between them on their common axis, when a current of 0.1 amp. is passed through them.

Solution:

The field is given by equation (1)

$$B = \frac{\mu_0 n\, IR^2}{2(R^2 + x^2)^{3/2}} \text{ Weber/metre}^2$$

The field due to both the coils is given by,

$$B = 2 \times \frac{\mu_0 n\, IR^2}{2(R^2 + x^2)^{3/2}}$$

Here $I = 0.1$ amp, $n = 60$, $\mu_0 = 4p \times 10^{-7}$ weber/amp. metre

$R = 7$ cm $= 0.07$ m, $x = 9$ cm $= 0.09$ m

therefore,

$$B = \frac{(4\pi \times 10^{-7}) \times 60 \times 0.1 \times (0.07)^2}{[(0.07)^2 + (0.09)^2]^{3/2}} \text{ Wb/m}^2$$

$$= 2.5 \times 10^{-5} \text{ weber/metre}^2.$$

Example 8:

If the magnetic field normal to the plane of a circular coil of n turns and radius r which carries a current I amp. is measured on the axis of the coil at a small distance h from the centre of the coil, show that it is smaller than at the centre by the fraction

$$\frac{3}{2}\frac{h^2}{r^2}.$$

Solution:

Using equation (1) the field at a distance h from the centre of the circular coil of radius r, is given by,

$$B = \frac{\mu_o n Ir^2}{2(r^2 + h^2)^{3/2}}$$

$$= \frac{\mu_o n Ir^2}{2r^3(1 + h^2/r^2)^{3/2}}$$

$$= \frac{\mu_o n I}{2r}\left(1 + \frac{h^2}{r^2}\right)^{-3/2}$$

Since h is very small as compared with r, therefore, using binomial theorem, we get,

$$B = \frac{\mu_o n I}{2r}\left(1 - \frac{3}{2}\frac{h^2}{r^2}\right)$$

at the centre h = 0.

Therefore, the magnetic field at the centre of the coil,

$$B_o = \frac{\mu_o n I}{2r}$$

The decrease in the magnetic field = $B_o - B$

$$= \frac{\mu_o n I}{2r} - \frac{\mu_o n I}{2r}\left(1 - \frac{3}{2}\frac{h^2}{r^2}\right)$$

$$= \frac{\mu_o n I}{2r} \times \frac{3}{2}\frac{h^2}{r^2}.$$

Therefore, the fraction of the decrease of magnetic field.

$$= \frac{B_o - B}{B_o} = \frac{3}{2}\frac{h^2}{r^2}$$

Example 9:

A long solenoid of length 1 metre and mean radius 10 cm consists of 1000 turns of wire. A current of 20 amps. flows through it. Calculate the magnetic field on its axis

(i) at its centre

(ii) at one of its end.

Solution:

The field at any point on the axis of a solenoid is given by equation (1).

$$B = \frac{\mu_0 n\, I}{2} (\cos \theta_1 \quad \cos \theta_2)$$

As given, $I = 20$ amp, $n = 1000$ turns/metre

$$\mu_o = 4\pi \times 10^{-7} \text{ weber/amp. metre}$$

(i) Referring we have at the centre O,

$$\cos \theta_1 = \frac{0.05}{[(0.5)^2 + (0.05)^2]^{1/2}}$$

$$= \frac{0.05}{[0.2525]^{1/2}}$$

and $\cos \theta_2 = - \cos (\pi - \theta_1) = - \dfrac{0.05}{[0.2525]^{1/2}}$

therefore,

$$B = \frac{(4\pi \times 10^{-7} \text{ weber / amp. metre}) \, (100 \text{ turns / m}) \times (20 \text{ amp})}{2} \times \frac{1}{(0.2525)^{1/2}}$$

$$= 2.5.2 \times 10^{-2} \text{ Weber/metre}^2$$

(ii) Referring we have at an end P,

$$\cos \theta_1 = \frac{1}{[(1)^2 + (0.05)^2]^{1/2}}$$

$$= \frac{1}{\sqrt{(1.0025)}}$$

$$\cos\theta_2 = \cos\pi/2 = 0$$

therefore, $$B = \frac{(4\pi\times10^{-7}\times1000\times20)}{2}\times\left(\frac{1}{\sqrt{(1.0025)}}-0\right)$$

$$= 1.256 \text{ weber/metre}^2$$

Example 10:

An electron has a velocity of 10^4 meters/sec. normal to a magnetic field of 0.1 weber/metre^2. Find the radius of the electron path and also it frequency.

Solution:

From equation (1), the radius,

$$R = \frac{9.1\times10^{-31}\times10^{4}}{1.6\times10^{-19}\times0.1} = 5.7\times10^{-7}\text{ metre}$$

This is a very small circle. The frequency.

$$n = \frac{v}{2\pi R} = \frac{10^4}{2\pi\times5.7\times10^7}$$

$$= 2.8\times10^9 \text{ rPs.}$$

Example 11:

Find the frequency of the oscillator that is to be connected between the dees of a cyclotron with a magnetic field $\vec{B}$ = 1.5 weber/metre^2 for accelerating deutron. How the result is modified if we choose a-particle or proton?

Solution:

$$q = 1.6\times10^{-19}\text{ couls.}$$

Mass of deutron = 2 × mass of proton

$$m = 3.34\times10^{-27}\text{ kg.}$$

From equation (1),

$$n = \frac{qB}{2\pi m}$$

$$= \frac{1.6\times10^{-19}\times1.5}{2\pi\times3.34\times10^{-27}} = 1.14\times10^{7}\text{ RPS}$$

$= 1.4$ MC/sec. (1 MC $= 10^6$ cycles)

Accordingly, the oscillator frequency must be 11.4 MC/sec. If instead of deutron proton is accelerated then,

n = 22.8 MC/sec. since q/m for proton = twice of deutron

and for a = particle n = 11.4 M/C sec. as q/m for a particle = q/m of deutron.

Example 12:

In the above example, if the radius of the dees is 50 cm. find the energy of the emergent deutrons. Also calculate the energy of a = particle or proton if they were accelerated under the same condition

Solution:

From equation (1),

$$W = \frac{1}{2}\frac{(RqB)^2}{m}$$

$$= \frac{(0.5\times1.6\times10^{-19}\times1.5)^2}{2\times3.34\times10^{-27}}$$

$$= 2.15 \times 10^{-12} \text{ joule}$$

$$= \frac{2.15\times10^{-12}}{1.6\times10^{-19}}\text{eV}$$

$$= 1.33 \times 10^7 \text{ eV}$$

$$= 1.33 \text{ MeV} \qquad (1 \text{ million electron volt} = 10^6 \text{ eV})$$

If either a = particles or protons have been used the energy would be double, *i.e.*, 26.6 MeV because for a = particle q in the above example will be 2q and m will be 2m.

Hence,

$$W = \frac{4}{2}\left(\frac{1}{2}\frac{(RqB)^2}{m}\right) = 26.6\text{MeV}$$

Similarly for proton q s same. But mass is half of the deutron, *i.e.*, q = q, m = m/2,

$$W = 2\left[\frac{1}{2}\frac{(RqB)^2}{m}\right] = 26.6\text{MeV}$$

Example 13:

A stream of protons and deutrons in a vacuum chamber enters a uniform magnetic field Both are subjected to the same accelerating potential, hence kinetic energies are the same. If ion stream is perpendicular to the magnetic field and protons move in a circular path of radius 15 cm, find the radius of the path traversed by the deutrons. Given that mass of a deutron is twice that of a proton.

Solution:

The radius is given by equation (1),

$$R = \frac{mv}{qB}$$

So that, radius of the path of proton

$$R_p = \frac{m_p v_p}{qB}$$

and radius of the path or deutron

$$R_d = \frac{m_d v_d}{qB}$$

Charge on deutron is same as on proton and moving in the same field. Thus,

$$\frac{R_d}{R_p} = \frac{m_d v_d}{m_p v_p}$$

Since they have same kinetic energies, *i.e.*,

$$\frac{1}{2} m_d \; v_d^2 = \frac{1}{2} \; m_p \; v_p^2$$

or

$$\frac{v_d}{v_p} = \sqrt{\frac{m_p}{m_d}}$$

Since $m_d = 2\ m_p$, so

$$\frac{v_d}{v_p} = \sqrt{\frac{m_p}{2m_p}} = \frac{1}{\sqrt{2}}$$

Hence,

$$\frac{R_d}{R_p} = \frac{2m_p}{m_p} \cdot \frac{1}{\sqrt{2}} = \sqrt{2}$$

or $\quad R_d = \sqrt{2} R_p$ since $R_p = 15$ cm (given)

hence $R_d = 0.15 \times \sqrt{2} = 0.212$ metre $= 21.2$ cm.

Example 14(a):

A cyclotron accelerating protons has dees of 40 cm. radius and a radio frequency supply of 12 megacycles per second at 10000 volts maximum value. Calculate.

(1) The magnetic field B

(2) The speed and kinetic energy acquired by protons

(3) Maximum number of revolutions and time spent by a proton within dees.

Solution:

(1) From equation (1),

Cyclotron frequency $n = \frac{qB}{2\pi m}$

or $$B = \frac{2\pi mn}{q}$$

$$= \frac{2 \times 3.14 \times 1.67 \times 10^{-27} \times 12 \times 10^{6}}{1.6 \times 10^{-19}}$$

$$= 0.79 \text{ weber/metre}^2$$

(2) From equation (2),

Velocity acquired by proton,

$$v = \frac{qBR}{m}$$

$$= \frac{1.6 \times 10^{-19} \times 0.79 \times 0.40}{1.67 \times 10^{-27}}$$

$$= 3 \times 10^{7} \text{ metre/sec.}$$

Energy $= 1/2\ mv^2 = 1/2 \times 1.6 \times 10^{-27} \times (3 \times 10^{7})^2$ joules

$= 7.5 \times 10^{-13}$ joules

$$= \frac{7.5 \times 10^{-13}}{1.6 \times 10^{-19}} \text{eV}$$

$$= 4.7 \times 10^{6} \text{ eV.}$$

Number of revolutions inside dees

$$N = \frac{WeV}{2qV} \times 1.6 \times 10^{-19}$$

$$N = \frac{4.7 \times 10^6 \times 1.6 \times 10^{-19}}{2 \times \times 1.6 \times 10^{-19} \times 10,000}$$

$$= 235$$

Time spent within dees $= 235/n$

$$= \frac{235}{12 \times 10^6}$$

$$= 195 \times 10^{-5} \text{ sec.}$$

Example 14(b):

A cathode-ray tube with electrostatics deflection has an accelerating voltage, V_a = 1500 volts, length of deflecting plate l = 1 cm. Spacing between the plates (d) = 1 cm and screen is placed at a distance L = 29.5 cm from the end of the deflecting plates. Find the voltage required to deflect the spot by 1 cm on the screen. Assuming that there is no fringing of the field.

Solution:

From eq. (1)

$$\frac{V_d}{Y} = \frac{2d\ V_a}{l(L + l/2)} = \frac{2 \times 10^{-2} \times 1500}{10^{-2} \times (29.5 + 0.5)10^{-2}}$$

or $$\frac{V_d}{Y} = 10,000 \text{ volts/metre}$$

or $$\frac{V_d}{Y} = 100 \text{ volts/cm}$$

The required voltage is 100 volts.

Example 15:

A cathode-ray tube with magnetic deflection has an accelerating voltage V = 1500 volts, a magnetic deflecting field is uniform over length l = 2 cm and screen is placed at a distance L = 29 cm from the end of the deflecting field. Find the magnetic field B required to deflect the spot of an electron beam 1 cm on the screen.

Solution:

From Eq. (1),

$$\frac{B}{Y} = \frac{1}{l(L+l/2)} \sqrt{\frac{2m\,V_a}{e}}$$

For an electron on substituting e and m, we get,

$$\frac{B}{Y} = \frac{1}{l(L+l/2)} \; 3.38 \times 10^6 \sqrt{V_a}$$

or $$\frac{B}{Y} = \frac{3.38 \times 10^6 \times 1500^{1/2}}{2 \times 10^{-2}(29+1)10^{-2}}$$

or $$\frac{B}{Y} = 2.18 \times 10^{-2} \text{ weber/metre}^2\text{/metre}$$

or $$\frac{B}{Y} = 2.18 \times 10^{-4} \text{ weber/metre}^2\text{/cm}$$

Therefore the required magnetic field is 2.18×10^{-4} weber/metre^2.

Example 16:

Calculate the value of the electric field intensity which will give a proton an acceleration equal to the acceleration due to gravity.

Solution:

$$\text{Force} = eE = mg$$

or $$E = mg/e$$

$$E = \frac{1.67 \times 10^{-27} \times 9.8}{1.6 \times 10^{-19}} \text{ volt / cm}$$

$$= 10.2 \times 10^{-19} \text{ volt/cm.}$$

Example 17:

Two parallel plates are kept in air 1 cm apart and potential difference of 100 volt is applied. If there be an electron resting on the plate at lower potential. Calculate (1) the force on the electron, (2) the kinetic energy of this electron, when it reaches the other plate.

Solution:

$$E = \frac{V}{d} \; \frac{100}{0.01} = 10^4 \text{ volt / cm.}$$

Force $$F = eE = 1.6 \times 10^{-19} \times 10^4 = 1.6 \times 10^{-15} \text{ Newton}$$

Acceleration

$$a = \frac{F}{m} = \frac{1.6 \times 10^{-15}}{9.1 \times 10^{-31}} = 1.75 \times 10^{15} \text{ metres / sec}^2.$$

velocity when it reaches the other plate,

$$v = Ö2ad \text{ Since } v^2 = u^2 + 2aS \text{ and}$$

$$u = 0,\ S = d$$

or $$v = \sqrt{2} \times 1.75 \times 10^{15} \times 0.01$$

$$= 5.9 \times 10^6 \text{ m/sec.}$$

Therefore, the kinetic energy

$$= \frac{1}{2} mv^2 = \frac{1}{2} \times 9.1 \times 10^{-31} \times \left(5.9 \times 10^6\right)^2$$

$$= 1.6 \times 10^{17} \text{ joules.}$$

Example 18:

What must be the initial velocity of electrons which can just strike a conductor at – 1200 volts potential.

Solution:

The electrons, having negative charge, are repelled by a conductor at negative potential. In order that an electron just reaches the conductor, it should have the kinetic energy given by,

$$\frac{1}{2} m v^2 = eV$$

$$v = \sqrt{\frac{2eV}{m}} = \sqrt{\frac{2 \times 1.6 \times 10^{-19} \times 1200}{9.1 \times 10^{-31}}}$$

$$= 2.06 \times 10^6 \text{ metre/sec.}$$

Example 19:

A positive ion of charge q and mass m is accelerated by a potential difference V, passes with a constant velocity through the space between two plates separated by distance d. Show that transit time between plates is $\frac{d}{(2qV/m)^{1/2}}$.

Solution:

From equation (1),

$$v = \sqrt{\frac{2qV}{m}}$$

If t be the transit line between the plates,

$$t = \frac{d}{v} = \frac{d}{\left(\frac{2qv}{m}\right)^{1/2}}$$

Example 20(a):

The electric field between the plates of a cathode-ray oscilloscope is 1.2 × 10⁴ Newton/coul. What deflection will an electron experience if it enters at right angles to the field with a kinetic energy of 2000 eV? The plates are 1.5 cm long.

Solution:

Using equation (1),

$$y = \frac{qE_y}{2m\,v_x^2}x^2$$

As given $1/2\ m\ v_x^2 = 2000$ eV

$$y = \frac{qE_y x^2}{4\times(2000\times1.6\times10^{-19})}$$

$$= \frac{1.6\times10^{-19}\text{ coul}\times1.2\times10^4\text{ Nt / coul}\times(1.5\times10^{-2}\text{ metre})^2}{4\times3.2\times10^{-16}\text{ joul}}$$

$$= 3.4 \times 10^4 \text{ metre}$$

$$= 0.34 \text{ mm}$$

Example 20(b):

An electron beam is accelerated through a potential difference of 1000 V. An electric field is applied at right angles to the path of the beam by means of rectangular parallel plates. The distance between the plates is 2.5 cm and potential difference 50 V. The edges of the plates are at a distance of 25 cm and 30 cm from the fluorescent screen. What is the displacement of the fluorescent spot ?

Solution:

$$E = \frac{50}{2.5\times10^{-2}}\ \frac{\text{Volts}}{\text{metre}}$$

Kinetic energy = $1/2\ mv^2 = eV$

$$1/2 \times 9.1 \times 10^{-31}\times v^2 = 1000 \times 1.6 \times 10^{-19}$$

On solving this for v, we get,

$$v = 5.6 \times 10^7 \text{ metre/sec}$$

Using equation (1).

$$y = \frac{qE}{m}\left(\frac{l}{2} + L\right)\frac{l}{v^2}$$

We have omitted the subscript and assumed that E is along y-axis and particle is moving along x-axis.

Substituting $L = 25 \times 10^{-2}$ metre

$l = 5 \times 10^{-2}$ metre

and v and E calculated above, we get,

$$y = \frac{50 \times 1.6 \times 10^{-19}}{2.5 \times 10^{-2} \times 9.1 \times 10^{-31}}\left[0.25 + \frac{0.05}{2}\right]\frac{5 \times 10^{-2}}{\left[5.65 \times 10^{7}\right]^2}$$

$$= 0.153 \times 10^{-2} \text{ metre}$$

$$= 0.153 \text{ cm.}$$

EXERCISES

1. What is Laplace rule? Derive from it Biot-Savart law. Obtain counter part of Coulomb' law of force between charges for two finite current elements.
2. Discuss the interaction between two parallel long currents. Define ampere using the expression you arrive at.
3. Obtain an expression for the magnetic field along the axis of circular coil carrying current.
4. Give the theory and working of Helmholtz galvanometer. How do you get uniform magnetic field between the coils?
5. A closely wound solenoid of 1200 turns has an axial length 80 cm and radius of 1.5 cm. A current of 1.2 amp flows in the solenoid. Find the flux density at the middle of the axis.
6. A long solenoid of length 1 metre and radius of cross-section 1.5 cm has five layers of windings of 850 turns each. If the solenoid carries a current 5.0 amp. calculate the values of

   (1) Magnetic induction and

   (2) Magnetic flue ϕ_B for a cross-section of the solenoid at the centre of the solenoid.

7. A solenoid of length 20 cm and radius 2 cm is wound uniformly with 3000 turns of wire. It carries a current of 2 amperes. What is the value of B?

   (1) on the axis of solenoid at the middle

   (2) on the axis at one end

   And what is flux ϕ_B

   (1) Through the coil at the middle and

   (2) Through one end.

8. If the coils of the Helmholtz galvanometer have diameter 14 cm and number of turns 275 each. Find the magnetic field midway between the coils when their distance apart is 7 cm and a current of 0.1 amp. flows through them.

9. State and prove Ampere's law. Apply it to calculate magnetic field due to

   (a) a solenoid and

   (b) a Toroid carrying current.

10. Discuss the phenomena, interaction between two parallel currents by Ampere's law.

11. Establish the relation,

$$\nabla \times \vec{B} = \mu_o \vec{j}$$

   and fully explain the meaning of $\nabla \cdot \vec{B} = 0$

12. What are vector and scalar magnetic potentials? How the idea of vector potential is introduced? Obtain an expression for magnetic vector potential due to an electric current flowing in a wire and hence calculate the magnetic field.

13. Discuss the equivalence of magnetic shell and circuit carrying electric current.

14. Two long parallel wires, each carrying a current of 10 amp. in the same direction are separated by 10 cm. Calculate the force between the wires per unit length.

15. A square loop of wire of edge 10 cm, carries a current of 10 amp. Calculate the value of B at the centre of the loop.

16. Two long straight wires are kept in free space at a distance of 1.0 metre and 1.0 amp current is flowing in the same direction

in each wire. Calculate the direction and magnitude of the mutual force per metre between the two wires. What will be the difference if the current flows in mutually opposite direction?

17. A current of 30 amps flows in a long wire, a rectangular loop ABCD carrying current 20 amp is situated 1.0 cm away from this wire. If the side of loop facing wire is 30 cm and other side is 8 cm. Calculate the magnitude and direction of the resultant force acting on the loop due to magnetic field of the current in the straight wire.

18. The axis of circular coil of radius 5 cm. having 10 turns and carrying a current of 5 amp. makes an angle q with a uniform magnetic field $\vec{B}$. Find the torque on the coil.

19. A current of 30 amp. flows in a flat circular coil having 100 closely wound turns of radius 10 cm. What is flux density at the centre of the coil?

20. What is Biot-Savart law? Apply it to determine the magnetic field.

    (1) Due to a long straight wire carrying current.

    (2) Due to a wire of finite length carrying current.

21. What is a solenoid? Obtain an expression for the field on the axis of solenoid.

23. State and prove the Ampere's circuital law *i.e.*,

$$\oint \vec{B}.d\vec{l} = \mu_0 I$$

Illustrate it by determining the magnetic field of a current carrying long wire.

4

Maxwell's Equations and Applications of Electromagnetic Induction

We have discussed so far the basic laws of electrostatics, magnetostatic and the effect of changing magnetic field. However, we still, do not know the basic laws of electricity and magnetism that apply to the general case–namely, the situation in which the variations in the electric and magnetic fields are so rapid, that we are not justified in assuming that they constitute small departure from the electrostatics and magnetostatics cases. This general case occurs in many practical situations.

MAXWELL'S EQUATIONS

The complete equations of electromagnetic theory are listed below:

$$\nabla.\vec{D} = r \qquad (a)$$

$$\nabla.\vec{B} = 0 \qquad (b)$$

$$\nabla \times \vec{E} = -\nabla.\vec{B}\frac{\delta \vec{B}}{\delta t} \qquad (c) \qquad ...(1)$$

$$\nabla \times \vec{B} = \mu_o \left(\vec{J} + \frac{\delta \vec{D}}{\delta t}\right) \qquad (d)$$

Maxwell equations were not generally accepted for many years after they were postulated (1873). *His curl equations involving* $\nabla \times \vec{E}$ *and* $\nabla \times \vec{B}$ *implied that time-varying electric and magnetic fields it free space* (empty space) *were interdependent, a changing electric field being able to generate a magnetic field, and vice versa.* The radio waves were unknown at the time and it was after 15 years (1888) when

Hertz demonstrated that electromagnetic waves were possible as predicted by Maxwell.

Along with Maxwell's equations certain other fundamental relations are of importance in dealing with electromagnetic problems. Among these may be mentioned Ohm's law at a point.

$$\vec{J} = \sigma\vec{E}, \qquad ...(2)$$

The continuity relation

$$\nabla.\vec{J} = -\frac{\delta\rho}{\delta t}, \qquad ...(3)$$

The force relations,

$$\vec{F} = q\vec{E}. \qquad ...(4)$$

$$d\vec{F} = Id\vec{l} \times \vec{B} \qquad ...(5)$$

and Lorentz equation,

$$\vec{F} = q\,(\vec{E} + \vec{v} \times \vec{B}), \qquad ...(6)$$

and the relation between $\vec{E}$ and $\vec{D}$ and between $\vec{B}$ and $\vec{H}$ as given by

$$\vec{D} = \varepsilon\vec{E} = \varepsilon_0\vec{E} + \vec{P}, \qquad ...(7)$$

$$\vec{B} = \mu\vec{H} = \mu_0\,(\vec{H} + \vec{M}) \qquad ...(8)$$

The equations (1) to (8) together with Newton's laws of motion are sufficient to explain any electromagnetic phenomena.

MAXWELL'S EQUATIONS IN FREE SPACE

In the preceding section, Maxwell's equations are stated in their general from. For free space, where the current density $\vec{J}$ and charge density ρ are zero as well as $\vec{D} = \varepsilon_0\vec{E}$ and $\vec{B} = \mu_0\vec{H}$, the equations reduce to a simple differential form,

$$\vec{\nabla}.E = 0 \qquad (a)$$

$$\nabla.\vec{B} = 0 \qquad (b)$$

$$\nabla \times \vec{E} = -\frac{\delta\vec{B}}{\delta t} \qquad (c) \qquad ...(1)$$

$$\nabla \times \vec{E} = \mu_0\varepsilon_0\frac{\delta\vec{E}}{\delta t} \qquad (d)$$

Any electromagnetic field must satisfy all of Maxwell equations. Let us assume that there is a changing charge and current distribution within a certain region of space, and let us consider the fields produced by this source of radiation. In the free space outside the region, the charge and current densities are everywhere zero.

In integral form the equations are

$$\oint_S \vec{E}.d\vec{S} = 0 \qquad \text{(a)}$$

$$\oint \vec{B}.d\vec{S} = 0 \qquad \text{(b)}$$

$$\oint \vec{E}.d\vec{l} = -\int_S \frac{\delta \vec{B}}{\delta t}.d\vec{S} \qquad \text{(c)} \qquad ...(2)$$

$$\oint \vec{B}.d\vec{l} = \mu_o \varepsilon_o \int_S \frac{\delta \vec{E}}{dt}.d\vec{S} \qquad \text{(d)}$$

First two equations are obtained by Gauss's divergence theorem ($\oint \vec{E}.d\vec{S} = \int_v \text{div } Edv$) and last two equations by Stores's theorem ($\oint \vec{B}.d\vec{l}$ $= \int_8 \text{curl } \vec{B}.d\vec{S}$).

ELECTROMAGNETIC WAVES IN FREE SPACE

For free space the charge density $\rho = 0$ and the current density $\vec{J} = 0$ and K (relative permittivity) = 1 and K_m (or μ_r) relative permeability = 1. Therefore, the constitutive equations are,

$$\vec{D} = \varepsilon_0 \vec{E}$$

$$\vec{B} = \mu_0 \vec{H}$$

$$J = 0$$

$$r = 0$$

Maxwell's equations are given by Eq. (1.20) *i.e.*,

$$\nabla.\vec{E} = 0 \qquad \text{(a)}$$

$$\nabla.\vec{B} = 0 \qquad \text{(b)}$$

$$\nabla \times \vec{E} = -\frac{\delta \vec{B}}{\delta t} \qquad \text{(c)}$$

$$\nabla \times \vec{B} = \epsilon_o \mu_o \frac{\delta \vec{E}}{dt} \qquad \text{(d)}$$

To obtain the differential equation of electromagnetic wave propagation we eliminate $\vec{B}$ from equations (1a) and (1b)

Taking the curl of equation (1c), *i.e.,*

$$\nabla \times \nabla \times \vec{E} = -\nabla \times \frac{\partial \vec{B}}{\partial t}$$

$$= -\frac{\delta}{\delta t}(\nabla \times \vec{B})$$

Substituting $\nabla \times \vec{B}$ from eq. (1.20d), we get,

$$\nabla \times \nabla \times \vec{E} = -\frac{\delta}{\delta t}\left(\varepsilon_o \mu_o \frac{\delta \vec{E}}{\delta t}\right)$$

$$= -\mu_o \varepsilon_o \frac{\delta^2 E}{\delta t^2}$$

According to vector product rule,

$$\nabla \times \nabla \times \vec{E} = \nabla(\nabla . \vec{E}) - \nabla^2 \vec{E} = -\nabla^2 \vec{E}$$

Since, $\nabla . \vec{E} = 0$, [by eq. (1(a)] hence above

relation reduces,

$$-\nabla^2 \vec{E} = -\mu_o \varepsilon_o \frac{\delta^2 \vec{E}}{\delta t^2}$$

or $$\nabla^2 \vec{E} = \mu_o \varepsilon_o \frac{\delta^2 E}{\delta t^2} \qquad ...(2)$$

Similarly $\vec{E}$ can be eliminated from Eqs. (1c) and (1d) yielding,

$$\nabla^2 \vec{B} = \mu_o \varepsilon_o \frac{\delta^2 E}{\delta t^2} \qquad ...(3)$$

Equations (2) and (3) representing the relation between the space and time variation of $\vec{E}$ and $\vec{B}$ are called "three dimensional wave equation" for $\vec{E}$ and $\vec{B}$ respectively. If we substitute,

$$\nabla^2 = \left(\frac{\delta^2}{\delta x^2} + \frac{\delta^2}{\delta y^2} + \frac{\delta^2}{\delta z^2}\right)$$

and $\vec{E} = \hat{i}E_x + \hat{j}E_y + \hat{k}E_z$

Then equation (1.22) becomes,

$$\left(\frac{\delta^2}{\delta x^2}+\frac{\delta^2}{\delta y^2}+\frac{\delta^2}{\delta z^2}\right)(\hat{i}E_x + \hat{j}E_y + \hat{k}E_z)$$

Similar expression for $\vec{B}$ is obtained if we analyse Eq. (3)

The wave equation given by Eqs. (2) and (3) is sometimes called "D" *Alembert's equation", have been integrated by him.*

Each component of $\vec{E}$ and $\vec{B}$ is identical with the familiar wave equation,

$$\frac{\delta^2 y}{\delta x^2} = \frac{1}{v^2}\frac{\delta^2 y}{\delta t^2}$$

The factor $\mu_o\varepsilon_o$ has the same significance as the $1/v^2$. Dimensionally eq. (2),

$$\frac{\text{volts}}{\text{met}^3} = \mu_o\varepsilon_o \frac{\text{volt}}{\text{met.sec.}^2}$$

since $\quad E = \text{volt/metre}$

or $\quad 1/\sqrt{\mu_o\varepsilon_o} = \text{met/sec.}$

Same result is obtained if we analyse Eq. (3).

Thus it appears that $1/\sqrt{\mu_o\varepsilon_o}$ has the dimensions of velocity and is in fact the velocity with which the wave propagates in the free space. Hence the velocity of wave in free-space $= 1/\sqrt{\mu_o\varepsilon_o}$

$$= \frac{1}{\sqrt{[4\pi\times 10^{-7}\ \text{weber / amp.met.}\ 8.85\times 10^{-22}\ \text{Coul}^2\ /\ \text{Nwt met}^2}}$$

Since 1 Weber $= \dfrac{\text{1 Newton. metre}}{\text{Amp.}}$

hence $\quad \dfrac{1}{\sqrt{\mu_o\varepsilon_o}}$

$$= \frac{1}{\sqrt{4\pi\times 10^{-10}\times 8.85\dfrac{\text{Newton. metre}}{\text{Amp}^2\text{.met.}}\cdot\dfrac{\text{Coul}^2}{\text{Newton} - \text{metre}^2}}}$$

$$= \frac{1}{\sqrt{4\times 3.14\times 8.85\times 10^{-19}\ \frac{1}{(\text{Coul}/\sec)^2}\cdot\frac{\text{Coul}^2}{\text{net}^2}}}$$

(Since Amp = Coul/Sec.)

$$= \frac{1}{\sqrt{4\times 3.14\times 8.85\times 10^{-19}}}\ \text{met./sec.}$$

$$= 3.00 \times 10^8 \ \text{met/sec.} \qquad ...(4)$$

which is same as the velocity of light in vacuum. Thus we can conclude that the field vectors $\vec{E}$ and $\vec{B}$ propagate as waves in vacuum (free space) with the velocity of light c. This result suggests that light waves are electromagnetic in nature.

ELECTROMAGNETIC WAVES IN ISOTROPIC NON-CONDUCTING MEDIA (*i.e.*, DIELECTRICS)

Maxwell's Equation

Let us first write the basic laws with medium of permittivity ε and permeability μ.

Gauss's law for electrostatics is given by,

$$\nabla.\vec{D} = \rho$$

Gauss's law for magnetic field

$$\nabla.\vec{B} = 0$$

Faraday's law, $\nabla \times \vec{E} = -\frac{\delta\vec{B}}{\partial t}$

Ampere's law in presence of medium is,

$$\oint \vec{H}.d\vec{l} = I$$

by Stokes's theorem is

$$\nabla \times \vec{H} = \vec{J}$$

But $\quad \vec{B} = \mu\vec{H}$

With this linear relation for $\vec{B}$ and $\vec{H}$, we can write Ampere's law in the presence of medium as,

$$\oint \frac{\vec{B}}{\mu} \cdot d\vec{l} = I$$

or
$$\oint \vec{B} . d\vec{l} = \mu I$$

Then by Stokes' theorem,

$$\nabla \times \vec{B} = \mu \vec{J}$$

Now with the introduction of displacement current $\frac{\delta \vec{D}}{\delta t}$ the Ampere's law is expressed by replacing J by $\vec{J} + \frac{\delta \vec{D}}{\delta t}$. Therefore, we can write,

$$\nabla \times \vec{B} = \mu \left(\vec{J} + \frac{\delta \vec{D}}{\delta t} \right)$$

Thus, the complete equations of electromagnetic theory are as follow:

$$\nabla . \vec{D} = 0 \quad \text{(a)}$$

$$\nabla . \vec{B} = 0 \quad \text{(b)}$$

$$\nabla \times \vec{E} = -\frac{\delta \vec{B}}{\delta t} \quad \text{(c)} \qquad ...(1)$$

$$\nabla \times \vec{B} = \mu \left(\vec{J} + \frac{\delta \vec{D}}{\delta t} \right) \quad \text{(d)}$$

with $\vec{D} = \varepsilon \vec{E}$

and $\vec{B} = \mu \vec{H}$

As dielectrics are non-conducting medium, therefore, current $\vec{J} = 0$ and in a homogeneous isotropic medium there is no volume distribution of charge, thus, the charge density $\rho = 0$. Hence the constitutive equations are

$$\vec{D} \text{ keo} \vec{E} = \varepsilon \vec{E}$$

$$\vec{B} = \text{momr} \vec{H} = \mu \vec{H}$$

$$\vec{J} = 0$$

$$\rho = 0$$

Hence the Maxwell's equations for dielectrics medium from equation (1).

$$\nabla . \vec{E} = 0 \qquad \text{(a)}$$

$$\nabla . \vec{B} = 0 \qquad \text{(b)}$$

$$\nabla \times \vec{E} = -\frac{\delta \vec{B}}{\delta t} \qquad \text{(c)} \qquad \ldots(2)$$

$$\nabla \times \vec{B} = \mu\varepsilon \frac{\delta \vec{E}}{\delta t} \qquad \text{(d)}$$

We can obtain the equation of propagation of wave in dielectrics medium by eliminating $\vec{E}$ from Eqns. (2c) and (2d) in the same manner as for free space.

Taking the curl of equation (2d) we have,

$$\nabla \times \nabla \times \vec{B} = \nabla \times \mu_o \frac{\delta \vec{E}}{\delta t}$$

As the medium is isotropic μ and ε remain constant throughout the material of the medium hence can be taken out of the ∇ operation.

Thus,

$$\nabla \times \nabla \times \vec{B} = \mu_o \nabla \times \frac{\delta \vec{E}}{\delta t}$$

$$= \mu\varepsilon \frac{\delta}{\delta t}(\nabla \times \vec{E})$$

Substituting,

$\nabla \times \vec{E}$ from eqn. (2c), we get,

$$\nabla \times \nabla \times \vec{B} \; \mu\varepsilon \frac{\delta}{\delta t}\left(-\frac{\delta \vec{B}}{\delta t}\right)$$

$$= -\mu\varepsilon \frac{\delta^2 \vec{B}}{\delta t^2}$$

But, $\nabla \times \nabla \times \vec{B} = \nabla (\nabla . \vec{B}) - \nabla^2 \vec{B}$ (see eqn. 2.52)

Since, $\nabla . \vec{B} = 0$

Therefore, we get,

$$\nabla \times \nabla \times \vec{B} = -\nabla^2 \vec{B}$$

Substituting in the above equation, we get,

$$-\nabla^2\vec{B} = -\mu\varepsilon\frac{\delta^2 B}{\delta t^2}$$

or
$$\nabla^2\vec{B} = \mu\varepsilon\frac{\delta^2\vec{B}}{\delta t^2} \qquad ...(3)$$

Similarly $\vec{B}$ can be eliminated from the equation (2c) and (2d) and we get,

$$\nabla^2\vec{E} = \mu\varepsilon\frac{\delta^2\vec{E}}{\delta t^2} \qquad ...(4)$$

Clearly equations (3) and (4) representing the relation between the space and time variation of magnetic field $\vec{B}$ and electric field $\vec{E}$ are called wave equations for $\vec{B}$ and $\vec{E}$ repetitively in a three dimensional case. If we substitute,

$$\nabla^2 = \frac{\delta^2}{\delta x^2}+\frac{\delta^2}{\delta y^2}+\frac{\delta^2}{\delta z^2}$$

$$\vec{B} = \hat{i}B_x + \hat{j}B_y + \hat{k}B_z$$

Then, $$\left(\frac{\delta^2}{\delta x^2}+\frac{\delta^2}{\delta y^2}+\frac{\delta^2}{\delta z^2}\right)(\hat{i}B_x + \hat{j}B_y + \hat{k}B_z)$$

$$= \mu\varepsilon\frac{\delta^2}{\delta t^2}(\hat{i}B_x + \hat{j}B_y + \hat{k}B_z)$$

Similar expression for $\vec{E}$ is obtained if we analyse eq. (4). Each component of $\vec{B}$ and $\vec{E}$ is identical with the familiar wave equation

$$\frac{\delta^2 y}{\delta x^2} = \frac{1}{v^2}\frac{\delta^2 y}{\delta t^2}$$

Here, therefore $\mu\varepsilon$ has the same significance as the $\frac{1}{v^2}$.

Dimensionally equation (3) is,

$$\frac{1}{\text{metre}^2}\cdot\frac{\text{weber}}{\text{metre}^2} = \mu\varepsilon\frac{1}{\text{sec}^2}\cdot\frac{\text{weber}}{\text{metre}^2}, \text{ since}$$

$$B = \frac{\text{Weber}}{\text{metre}^2} \text{ so that } \frac{1}{\mu\in} = \frac{\text{meter}}{\text{sec.}}$$

Same result is obtained if we analyse equation (4).

Thus, it appears that $1/\sqrt{\mu\varepsilon}$ *as the dimensions of velocity and is in fact the velocity with which the wave propagates in the dielectrics,* if represented by v, then

$$v = \frac{1}{\sqrt{\mu\varepsilon}} \qquad ...(5)$$

Thus we can conclude that the field vectors $\vec{B}$ and $\vec{E}$ propagates as waves with velocity $1/\sqrt{\mu\varepsilon}$. The velocity is a characteristic of the medium depending on the constant μ and ε.

INDEX OF REFRACTION

Since $\varepsilon = \varepsilon_o k$

and $\mu = \mu_o \mu_r$

where k is relative permittivity and μ_r is relative permeability

$$\text{Then, } v = \frac{1}{\sqrt{\mu\varepsilon}} = \frac{1}{\sqrt{\mu_o \mu_o \varepsilon_o k}}$$

$$= \frac{1}{\sqrt{\mu_o \varepsilon_o} \cdot \sqrt{\mu_r k}}$$

by equation (1), $1/\sqrt{\mu_o \varepsilon_o} = C$ (velocity of light), hence,

$$v = \frac{C}{\sqrt{\mu_r k}}$$

$$= \frac{C}{\eta} \qquad ...(2)$$

where η is the refractive index of the dielectric which is always greater than unity. *Thus velocity of a wave in a dielectric is less than in free space.* For non-magnetic (or non-ferrous) medium, permeability $\mu_r = 1$ refractive index,

$$h = \frac{C}{v} = \sqrt{\mu_r k} = \sqrt{k} \qquad ...(3)$$

SOLUTIONS OF THE WAVE EQUATIONS

Let us now discuss the solutions of wave equations (1) and (2). There are many solutions corresponding to different kinds of electromagnetic waves such as plane, spherical and cylindrical waves. We shall consider only uniform plane wave. Therefore, we shall assume that $\vec{E}$ and $\vec{B}$ are functions of only one space coordinate along which the wave travels and of the time t and there is no variation in the component of $\vec{E}$ and $\vec{B}$ in any other direction. If $\vec{E}$ and $\vec{B}$ vary along x-axis, then

$$\vec{E} = \vec{E}\,(x, t) \text{ and } \vec{B} = \vec{B}\,(x, t)$$

and all derivatives of the components of $\vec{E}$ and $\vec{B}$ with respect to y and z are zero. Thus,

$$\frac{\delta E_y}{\delta y} = 0, \quad \frac{\delta E_z}{\delta z} = 0, \quad \frac{\delta B_y}{\delta y} = 0 \text{ and } \frac{\delta B_z}{\delta z} = 0 \qquad ...(3)$$

Then eqs. (1) and (2), *i.e.*,

$$\nabla.\vec{E} = 0 \text{ and } \nabla.\vec{B} = 0$$

written in cartesian coordinates are,

$$\nabla.\vec{E} = \frac{\delta E_x}{\delta x} + \frac{\delta E_y}{\delta y} + \frac{\delta E_z}{\delta z} = 0$$

$$\nabla.\vec{B} = \frac{\delta B_x}{\delta x} + \frac{\delta B_y}{\delta y} + \frac{\delta B_z}{\delta z} = 0$$

for plane wave by equation (3),

$$\frac{\delta E_x}{\delta x} = 0 \qquad ...(4a)$$

$$\frac{\delta B_x}{\delta x} = 0 \qquad ...(4b)$$

Equations (4a) and (4b) indicate that E_x and B_x cannot be a function of x which is the direction of propagation. This means E_x and B_x are constant in x-direction. However, we are not interested in uniform fields since these are not the part of any wave equation, therefore, we shall set,

$$E_x = 0 \text{ and } B_x = 0 \qquad ...(4c)$$

Hence the vector $\vec{E}$ and $\vec{B}$ can have only y and z components *i.e., it has no "longitudinal" component but has only "transverse" components. Thus,*

$$\vec{E} = \hat{j}E_y + \hat{k}E_z$$

$$\vec{B} = \hat{j}B_y + \hat{k}B_z \qquad ...(4d)$$

To be more clear, the Maxwell's electromagnetic waves are purely transverse and never longitudinal. Also from equations (1.26c), we have,

$$\nabla \times \vec{E} = -\frac{\delta\vec{B}}{\delta t}$$

$$\begin{vmatrix} \hat{i} & \hat{j} & \hat{k} \\ \frac{\delta}{\delta x} & \frac{\delta}{\delta y} & \frac{\delta}{\delta z} \\ E_x & E_y & E_z \end{vmatrix} = -\left(\hat{i}\frac{\delta B_x}{\delta t} + \hat{j}\frac{\delta B_y}{\delta t} + \hat{k}\frac{\delta B_z}{\delta t}\right)$$

On equating the components separately, we get,

$$\frac{\delta E_z}{\delta y} - \frac{\delta E_y}{\delta z} = -\frac{\delta B_x}{\delta t}$$

$$\frac{\delta E_x}{\delta z} - \frac{\delta E_z}{\delta x} = -\frac{\delta B_y}{\delta t}$$

$$\frac{\delta E_y}{\delta z} - \frac{\delta E_x}{\delta y} = -\frac{\delta B_z}{\delta t}$$

Now remembering that $E_x = 0$ and $\frac{\delta}{\delta y} = \frac{\delta}{\delta z} = 0$, we get,

$$\frac{\delta B_x}{\delta t} = 0 \qquad ...(43e)$$

$$\frac{\delta E_z}{\delta x} = \frac{\delta B_y}{\delta t} \qquad ...(4f)$$

$$\frac{\delta E_y}{\delta x} = -\frac{\delta B_z}{\delta t} \qquad ...(4g)$$

Similarly from eq. (3d), *i.e.*,

$$\nabla \times \vec{B} = \mu\varepsilon \frac{\delta\vec{E}}{\delta t}$$

can be written in components, *i.e.*,

$$\frac{\delta B_z}{\delta y} - \frac{\delta B_y}{\delta z} = \mu\varepsilon \frac{\delta E_x}{\delta t}$$

$$\frac{\delta B_x}{\delta z} - \frac{\delta B_z}{\delta x} = \mu\varepsilon \frac{\delta E_y}{\delta t}$$

$$\frac{\delta B_y}{\delta x} - \frac{\delta B_x}{\delta y} = \mu\varepsilon \frac{\delta E_z}{\delta t}$$

Again remembering that $B_x = 0$ and $\frac{\delta}{\delta y} = \frac{\delta}{\delta z} = 0$, we get,

$$\frac{\delta E_x}{\delta t} = 0 \quad ...(4h)$$

$$\frac{\delta B_z}{\delta x} = -\mu\varepsilon \frac{\delta E_y}{\delta t} \quad ...(4i)$$

$$\frac{\delta B_y}{\delta x} = \mu\varepsilon \frac{\delta B_z}{\delta t} \quad ...(4j)$$

The equations [3(f, g, i, j)] clearly represent that E_z is associated with B_y and E_y is associated with B_z. Thus E_z and B_y are quite independent of E_y and B_z. Thus we can write,

$$\frac{\delta E_y}{\delta x} = -\frac{\delta B_z}{\delta t} \quad ...(4k)$$

and $$\frac{\delta B_z}{\delta x} = -\mu\varepsilon \frac{\delta E_y}{\delta t} \quad ...(4l)$$

Differentiating equation (4k) with respect to x and eqn. (3) with respect to t, we get

$$\frac{\delta^2 E_y}{\delta x^2} = -\frac{\delta^2 B_z}{\delta x \delta t}$$

and $$\frac{\delta^2 B_z}{\delta x \delta t} = -\mu\varepsilon \frac{\delta^2 E_y}{\delta t^2}$$

hence, $$\frac{\delta^2 E_y}{\delta x^2} = \mu\varepsilon \frac{\delta^2 E_y}{\delta t^2} \quad ...(5a)$$

or $$\frac{\delta^2 E_y}{\delta t^2} = \frac{1}{\mu\varepsilon} \frac{\delta^2 E_y}{\delta x^2} \quad ...(5b)$$

The velocity of wave $v = 1/\sqrt{\mu\varepsilon}$

The wave equation (3a) is a linear partial differential equation of the second order disrobing the propagation of the y-component of the electric field $\vec{E}$ along the x-axis with constant velocity v.

For simplifying the calculations let us choose our coordinate axes such that y-axis is parallel to the vector $\vec{E}$. We can do this since E_y and B_z are quite independent of E_z and B_y. Thus we suppose that $E_z = 0$ and $B_y = 0$, then

$$\vec{E} = \hat{j}E_y \text{ and } \vec{B} = \hat{k}B_z$$

Here, as we are interested in electromagnetic waves in which the field is simply periodic function of time *i.e.,* in general the time dependence $\vec{E}$ and $\vec{B}$ is $e^{j\omega t}$ for single fixed frequency.

Thus we write in equation (5a),

$$E_y(x, t) = f(x)\, e^{j\omega t} \qquad ...(6)$$

After substitution in (5a), we get,

$$\frac{d^2f}{dx^2} = -\frac{\omega^2}{v^2}f \qquad ...(7a)$$

This Equation has two solutions,

$$f(x) = e^{j\omega x/v}$$

and $f(x) = e^{-j\omega x/v}$

Hence a general solution for (3) is

$$E_y = E_{oy}\left[e^{j\omega(t+x/v)} + e^{j\omega(t-x/v)}\right] \qquad ...(7b)$$

Therefore, either term alone is a solution or the sum as in eqn. (7b) is a solution. We can choose 1st term of equation (7b) and take the imaginary part as solution then we get,

$$E_y = E_{oy} \sin \omega\left(t + \frac{x}{v}\right) \qquad ...(8)$$

The significance of eqn. (8) can be illustrated by evaluating E_y as a function of x for several values of time t

At $t = 0$, $E_y = \sin \omega \dfrac{x}{v}$, hence at $x = 0$, $E_y = 0$

At $t = \dfrac{T}{4}$, $\omega t = \dfrac{2\pi}{T}\dfrac{T}{4} = \dfrac{\pi}{2}$ and $E_y = \sin\left(\dfrac{\pi}{2} + \dfrac{\omega x}{v}\right)$

Hence at x = 0, $E_y = +1$, The maximum value of E_y is unity.

At $t = \frac{T}{2}$, $\omega t = \frac{2\pi}{T}\cdot\frac{T}{2} = \pi$ and $E_y = \sin\left(\pi + \frac{\omega x}{v}\right)$

Hence at x = 0, $E_y = -1$

The curves are shown in Fig. 4.1(a)

Focussing our attention on the crest of one of the waves as indicated by the point P, we note that as time progresses, P moves to the left. Thus we can conclude that equation (8) is representing a wave travelling to the left *i.e.,* negative x-axis. The point P is a point of constant phase and is characterised by the condition x + vt = const., taking the time derivatives $\frac{dx}{dt} + v = 0$

If we take, $E_y = E_{oy} \sin \omega\left(t - \frac{x}{v}\right)$

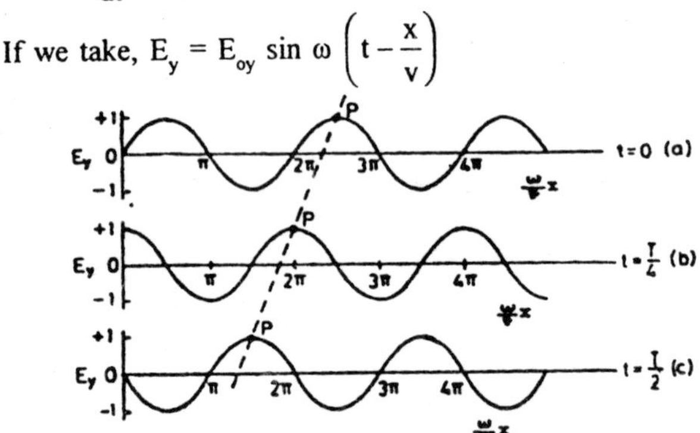

Fig. 4.1(a)

Then proceeding in the same manner we found that it represents the wave travelling to the right as shown in Fig. 4.1(b) positive x-axis. To "summaries" the negative sign in t ±x/y is associated with a wave to the right, while a positive sign is associated with a wave to the left. Other equivalent solution can also be used such as,

$$E_y = E_{oy} \cos \omega(t - x/v) \text{ or } E_y = E_{oy} \cos\left(\frac{\omega}{v}x - \omega t\right)$$

Again for simplifying the calculations, let us consider only the wave travelling in the positive direction of the x-axis, therefore, we have,

$$E_y = E_{oy} \sin \omega(t - x/v) \qquad \text{...(9a)}$$

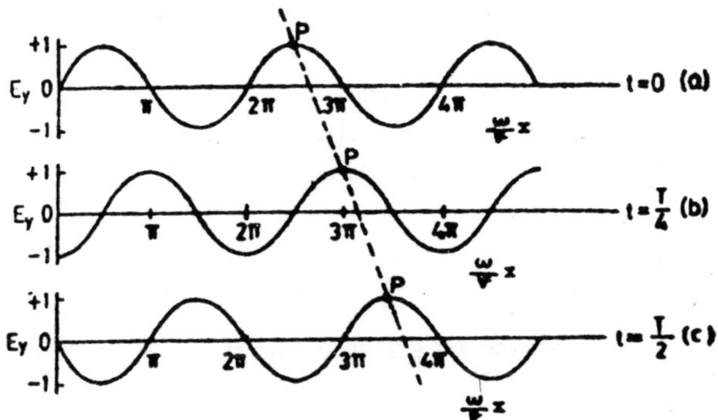

Fig. 4.1(b)

Now we can calculate B_z by substituting E_y in eq. (6k)

$$\frac{\delta}{\delta x}\left[E_{oy} \sin \omega(t - x/v)\right] = -\frac{\delta}{\delta t} B_z$$

$$-E_{oy} \cos \omega\left(t - \frac{x}{v}\right) \cdot \frac{\omega}{v} = -\frac{\delta}{\delta t} B_z$$

Integrating with respect to t, we get,

$$E_{oy} \sin \omega(t - x/v) \cdot \frac{1}{v} = B_z$$

or $$B_z = B_{oz} \sin \omega(t - x/v) \qquad \text{...(9b)}$$

where $B_{oz} = E_{oy}/v$ ...(10)

Hence ratio of eqns. (9a) and (9b)

$$\frac{E_y}{B_z} = \frac{E_{oy}}{B_{oz}} = v$$

or $$\frac{E_y}{B_z} = \frac{1}{\sqrt{\mu\varepsilon}} \qquad \text{...(11a)}$$

If the medium is free space then,

$$\frac{E_y}{B_z} = \frac{1}{\sqrt{\mu_o \varepsilon_o}} \qquad \text{...(11b)}$$

The value of E_y/B_z is real and depends on the medium in which wave is travelling. Equations (9a) and (9b) represent linearly polarised place harmonic wave travelling along the +x axis in which $\vec{E}$ is everywhere parallel to y-axis and $\vec{B}$ is parallel to z-axis. As the ratio E_y/B_y is a real positive constant, hence the two vectors $\vec{E}$ and $\vec{B}$ are in phase when $\vec{E}$ has its maximum value directed along +y direction, $\vec{B}$ has its maximum value directed along +z-axis.

By equation (11a)

$$B_z = \sqrt{\mu \in} E_y$$

Now multiplying the left side by $\hat{k}$ and right side by its equivalent $\hat{i} \times \hat{j}$, we get,

$$\hat{k} B_z = \sqrt{\mu\varepsilon} . \hat{i} \times \hat{j} \; E_y$$

or $$\vec{B} = \sqrt{\mu\varepsilon} \; \hat{i} \times \vec{E} \qquad \text{...(12a)}$$

Similarly, $E_y = \dfrac{1}{\sqrt{\mu\varepsilon}} B_z$

Multiplying left side by $\hat{j}$ and right side its equivalent $\hat{k} \times \hat{i}$, we get,

$$\hat{j} E_y = \frac{1}{\sqrt{\mu\varepsilon}} \; \hat{k} \times \hat{i} \, B_z$$

$$= -\frac{1}{\sqrt{\mu\varepsilon}} \; \hat{i} \times \hat{k} \, B_z$$

$$\vec{E} = -\frac{1}{\sqrt{\mu\varepsilon}} \; \hat{i} \times \vec{B} \qquad \text{...(12b)}$$

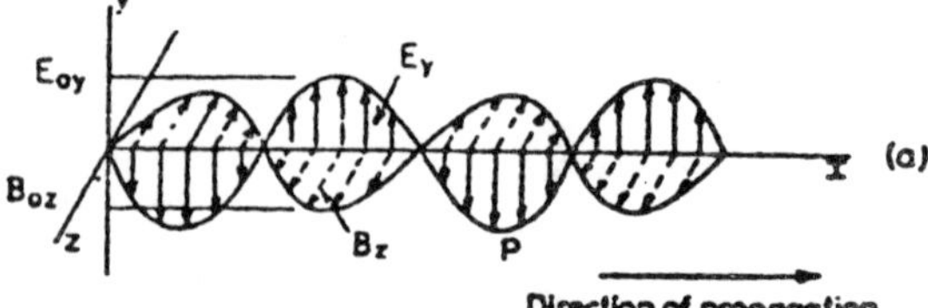

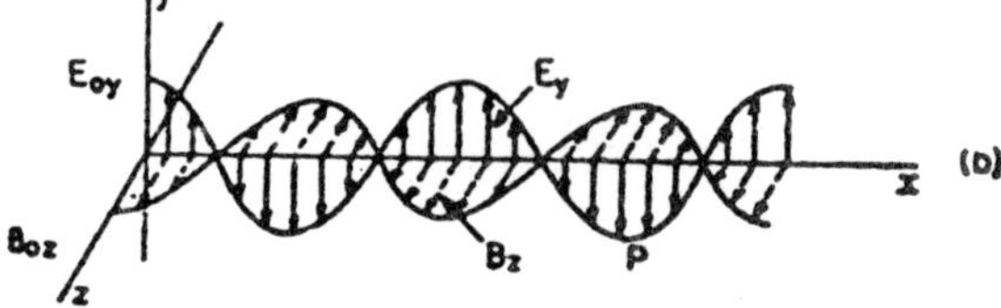

Fig. 4.2

The vectors $\vec{E}$ and $\vec{B}$ are perpendicular to one another, the directions of $\vec{E}$ and $\vec{B}$ are related to the direction of propagation by the rule of "right hand screw". The vectors $\vec{E}$ and $\vec{B}$ are such oriented that their vector product $\vec{E} \times \vec{B}$ points in the direction of propagation of the wave as shown in Fig. 4.2.

IMPEDANCE OF DIELECTRIC MEDIA

We can also write equation (1) in a more appropriate form as follows. Since $B_z = \mu H_z$ where μ is permeability of the medium,

$$\frac{E_y}{H_z} = \frac{\mu}{\sqrt{\mu\varepsilon}}$$

or
$$\frac{E_y}{H_z} = \sqrt{\mu / \varepsilon} \qquad ...(1)$$

However, for free space,

$$\frac{E_y}{H_z} = \sqrt{\frac{\mu_o}{\varepsilon_o}} \qquad ...(2)$$

Dimensions of equation (2) expressed in MKS units are,

$$\frac{\text{volts / metre}}{\text{Amps. / metre}} = \frac{\text{volts / metre}}{\text{Amps. / metre}} = \text{ohms}$$

Thus $\sqrt{\mu / \varepsilon}$ *has the dimensions of an impedance and the ratio* E_y/H_z *is called the impedance of the medium. For free space.* E_y/H_z = 376 ohms.

GRASSOT'S FLUX METER

It is essentially a moving coil galvanometer in which mechanical damping and restoring torque are very much reduced but electromagnetic damping is relatively increased. Hence it is a modified moving coil ballistic galvanometer and is used to measure magnetic flux. The advantage of a Grassot's fluxmetre over a ballistic galvanometer is that while measuring the flux, the change in flux did not take place rapidly, since the fluxmetre coil is at rest before and after the change. Thus is Grassot's fluxmetre, the deflection depends only upon the change of flux and not upon the time taken for the flux to change, so the deflection is independent of the rate at with flux changes. The moving coil is suspended by a single silk fibre, the upper end of the fibre being attached to a flat spiral spring A. The coil is connected through two spirals B. B of thin silver foil and is suspended in the magnetic field of a permanent magnet NS. The search coil is connected to the terminals T_1 and T_2 of the moving coil of the fluxmetre as shown in Fig. 4.3.

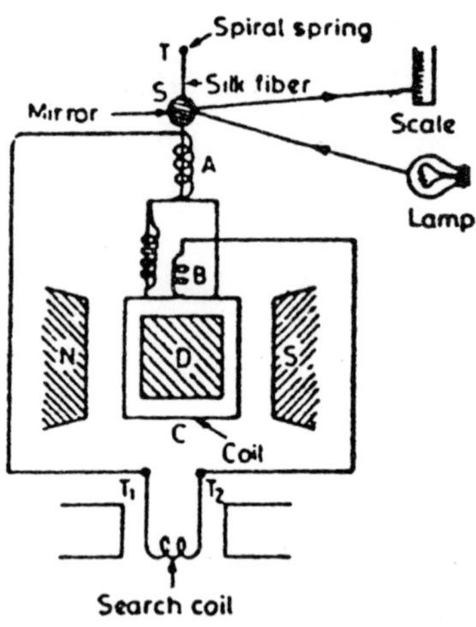

Fig. 4.3

The deflection of the coil is indicated by the usual lamp and scale arrangement. The search coil is inserted in the space where the magnetic flux is to be measured and then withdrawn, an emf is induced in the circuit which gives rise to a current. The corresponding deflection produced in the fluxmetre is noted.

The induced current = x/R

where x is induced emf and R is the resistance of the circuit

Suppose area of the fluxmetre coil	= A
Number of turns	= N
Magnetic field	= B
The magnitude of the deflecting couple	= iANB

(Since $\tau = \vec{m} \times \vec{B}$ and m = iA)

This couple acts on the fluxmetre coil. If i is the moment in inertia of the fluxmetre coil and $\frac{d\omega}{dt}$ is its angular acceleration then we have,

$$NBiA = I \frac{d\omega}{dt}$$

$$ABN \frac{\xi}{R} = I \frac{d\omega}{dt} \quad ...(1)$$

The emf ξ is the resultant of two emfs one due to the rate of change of flux *i.e.*, $\frac{d\phi}{dt}$ in the search coil and the other due to the rotation of fluxmetre coil with angular velocity ω *i.e.*, NABω

$$\xi = \frac{d\phi}{dt} - NAB\,\omega$$

Substituting ξ in equation (3.15), we get,

$$\frac{ABN}{R}\left(\frac{d\phi}{dt} - NAB\,\omega\right) = I \frac{d\omega}{dt}$$

Integrating and putting $\omega = \frac{d\theta}{dt}$

$$\frac{NAB}{R}\left(\int_0^t \frac{d\phi}{dt}\,dt - NAB\int_0^t \frac{d\theta}{dt}dt\right) = I\int_0^t \frac{d\omega}{dt}dt$$

The integral of the right-hand side *i.e.*,

$$\int_0^t d\omega = |\omega|_{t=0}^{t=t} = 0$$

because the coil is at rest before and after the change of flux *i.e.*, angular velocity is zero when t = 0 and t = t.

$$\frac{NAB}{R}\left[\int_0^t d\phi - NAB\int_0^t d\theta\right] = 0$$

or $$\int_0^t d\phi - NAB\int_0^t d\theta = 0$$

or $$\phi - NAB\phi = 0$$

or $$\phi = NAB\,\theta$$

or $$\phi = K\,\theta \quad ...(2)$$

where $K = NAB$.

Thus, *the total change in flux in the search coil is independent of time in which the change of flux takes place.* The deflection θ in the fluxmetre is proportional to the flux linkage in the search coil. Therefore, the displacement of the fluxmetre coil, when the search coil is inserted in or withdrawn from the unknown field, measures the flux linked with the search coil.

A.C GENERATOR OR DYNAMO

Principle

It is based on the principle of electromagnetic induction. When a coil rotates in a uniform magnetic field, the associated flux changes and an induced emf.is generated according to Faraday's law of electromagnetic induction. *Thus the rotation of coil (Mechanical Processes) produces induced emf (electrical effects) if the process is made continuous, then we can convert continuously mechanical energy into electrical energy.*

Suppose a coil of wire is rotated about an axis OO' in a uniform magnetic field $\vec{B}$. The flux passing through the coil.

$$\phi = \int_S \vec{B}.\hat{n}dS$$

where $\hat{n}$ is a vector along the outward normal to the plane of the coil and dS is an element of the area enclosed by the coil. Let at any instant θ be the angle between normal to the plane of the coil and $\vec{B}$. Further, let N be the total number of turns in the coil and A be the area of each turn, total flux

$$\phi = \int_S B \cos\theta \; dS$$

$$= B\cos\theta \int_S dS$$

$$= NAB\cos\theta, \text{ since } \int_S dS = NA$$

The induced emf in the coil at any instant t is given by,

$$E = -\frac{d\phi}{dt} = -\frac{d}{dt}(BAN\cos\theta)$$

$$= BAN\sin\theta\frac{d\theta}{dt}$$

If ω is the angular velocity, then ωt = θ, therefore,

$$E = BAN\,\omega\sin\omega t \quad ...(1)$$

The value of induced emf is maximum when $\omega t = \frac{\pi}{2}$, $\sin \omega t = 1$, *i.e.*, when the plane of the coil is parallel to the magnetic field, this maximum value of E is denoted by E_o.

$$E_o = BAN\omega$$

Equation (1) becomes,

$$E = E_o \sin \omega t \qquad ...(2)$$

This relation (2) shows the variation of emf with time.

1. When $\omega t = 0$, *i.e.*, the plane of the coil is perpendicular to the field (coil is vertical), $E = 0$.
2. When $\omega t = \frac{\pi}{2}$, *i.e.*, the plane of the coil is parallel to B (coil is horizontal) we have $E = E_o$.
3. When $\omega t = \pi$, *i.e.*, the plane of the coil is again perpendicular to the field, $E = 0$.
4. When $\omega t = \frac{3\pi}{2}$, *i.e.*, the plane of the coil is again parallel to the field $\sin\left(\pi + \frac{\pi}{2}\right) = \sin\frac{\pi}{2} = -1$, so that $E = -E_o$.

   It shows that the direction of the induced emf is again maximum but in the reverse direction.
5. When $\omega t = 2\pi$, the plane of the coil is again perpendicular to the field and $E = 0$.

Thus, the emf induced in the coil goes through a cycle. Such emf is called sinusoidal just like sine wave. It is also referred to as alternating emf. The corresponding current in the external circuit,

$$I = \frac{E}{R} = \frac{E_o}{R} \sin \omega t$$

$$= I_o \sin \omega t \qquad ...(3)$$

where I_o is maximum current.

Construction

A simple electric generator consists of a coil which can rotate in magnetic field. The coil of the wire is called the armature. It is held between the pole-pieces of strong magnet called the field magnet. When the armature rotates, the magnetic flux linked with it changes and the current is induced in the coil. This current is taken out by means of an arrangement called slip rings and brushes. The two rings are insulated from each other the brush lightly touches each ring and connects the armature with the external circuit.

Source of Energy

The emf is generated when the armature rotates, hence, there must by some device for rotating the armature. Such as in a cycle dynamo the armature is rotated by the rotation of the wheel. For other purposes, steam engines, gas turbines etc. are used to rotate the armature. Thus the mechanical energy is converted into electrical energy. *In every ;generator, some form of mechanical energy is put into the generator, and transformed into electrical energy. Therefore the generator basically does not produce energy but merely it transforms mechanical to electrical forms.*

D.C. GENERATOR OR DYNAMO

IN d.c. dynamo the current and emf are unidirectional. The principle of working is same as that of a.c. generator except that an arrangement is provided which reverses the emf after half cycle in a.c. generator.

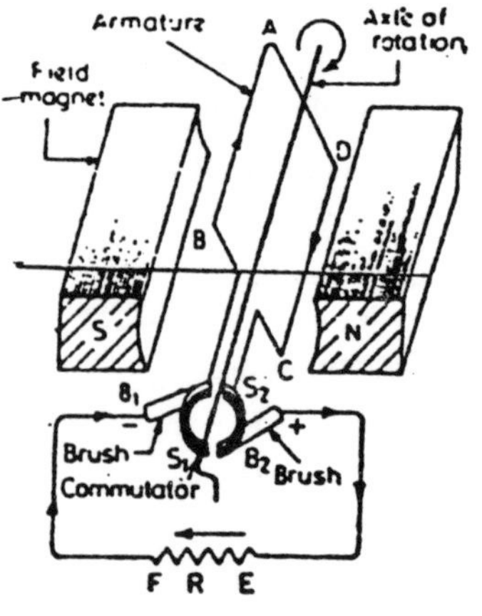

Fig. 4.4

This is achieved by reversing the connections to the ends of the coil of the generator, when the direction of emf changes every time. This is done by using a *split ring commutator* instead of the slip rings. The split ring consists of two segments S_1 and S_2 of a brass cylinder which are insulated from each other. Each end of the armature coil is connected to one segment of the split ring. *The brushes and the split rings are so mounted that the contact of one brush with the first segment changes over to that with the second segment as soon as the direction of emf in the coil* reverses. This is done by mounting the commutator on the same axis as the coil and rotating along with it. The emf reverses the sign when the plane of the coil is at right angles to the magnetic field $\vec{B}$ at this instant emf is zero. Such a split ring commutator is shown in Fig. 4.4.

The emf time graph is shown in Fig. 4.5.

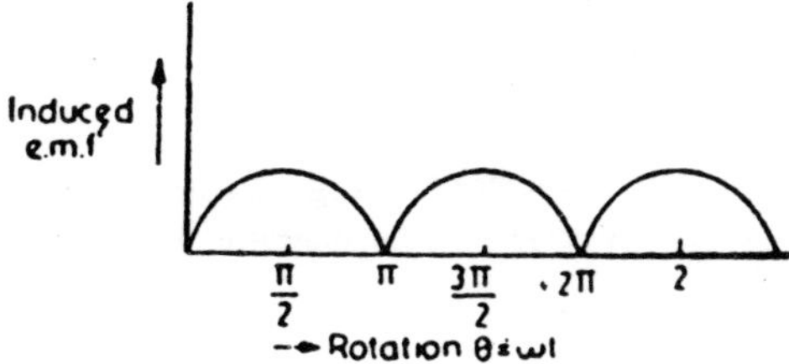

Fig. 4.5

ENERGY LOSSES IN GENERATOR AND THEIR REDUCTION

The energy losses in a dynamo are due to:

1. *Heating or copper losses.* Part of energy is wasted in overcoming the ohmic resistance of the armature coils, field coils and due to sparking at the brushes. This can be reduced by using thick copper wire of low resistance.
2. *Hysteresis Loss.* This is the energy used up in magnetising the core. This is reduced by choosing the material of the core having very thin B.H loop such as soft iron.
3. *Eddy Currents.* This is also known as iron loss and is due to eddy currents in the iron core. This is reduced by using laminated core.
4. *Flux Leakage.* Leakage of flux is reduced by making the pole-shoe large so as to enclose completely the armature and also by reducing the air gap.
5. *Mechanical Loss.* This is due to friction at the bearings, air resistance, etc. It is reduced by using roller or ball bearing and proper lubrication.

EFFICIENCY OF A GENERATOR

Efficiency is defined as

$$\eta = \frac{\text{Electrical energy supplied by the generator}}{\text{Mechanical energy supplied to the generator}}$$

If V is terminal voltage developed and I is current in external circuit, then power developed = VI. If the total power loss is W watts, then power used

$$= (VI + W) \text{ watts}$$

and

$$\%\eta = \frac{IV}{VI + W} \times 100 \qquad ...(1)$$

D.C. MOTORS

It is a device by which electrical energy is converted into mechanical energy. Hence, it is a reverse of the generator or dynamo.

Principle

The principle of a d.c. motor is the same as that of moving coil galvanometer which is the reverse of the d.c. generator. When a current carrying conductor capable of a free movement is placed in a magnetic field as shown in Fig. 4.6, it experiences a mechanical force and begins to move. When a coil carrying a current I is placed in a magnetic field $\vec{B}$, then it experiences a force $(\vec{F} = I\vec{l} \times \vec{B})$ which turns it about in a direction perpendicular to both the field and the current. Thus the coil rotates and the rotating torque

$$\tau = \vec{m} \times \vec{B}$$

where m is magnetic moment given by m = NIA, here A is area of the each coil and N are the number of the turns. When the plane of the coil is vertical, no torque acts on the armature (or coil). As the coil has sufficient inertia and frictional forces small it will cross the zero torque position. In order to make the motion of the coil continuous, the direction of the current is reversed after the vertical position (*i.e.*, after each half revolution) with the help of split ring commutator so the torque acts in the same direction. Thus, we see that after each half cycle, there are positions of no torque and the principle used is reverse of d.c. generator.

For smooth motion, the single coil is replaced by several coils set such as drum winding called armature.

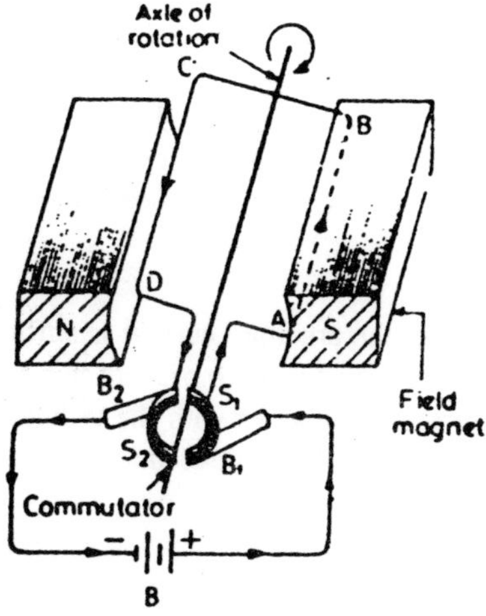

Fig. 4.6

The magnetic field is generally provided by electromagnets called field magnet. From the direct current sources, the armature is connected in series with electromagnet. Then motor is said to series-wound. If the armature is connected in parallel with the field magnet, the motor is then called parallel or shunt wound. Sometimes the windings over the armature and the field magnet are partly in series or partly in parallel, the motor is then called compound wound.

Back emf in a Motor

When the armature coil rotate an induced emf is set up in the coils of the winding and the direction of the induced emf opposes the current that rotates the armature. This emf is called back emf of the motor. Thus, every motor acts as a generator also.

The value of back emf goes on increasing steadily as the motor picks up speed. An equilibrium is established when this back emf plus the voltage drop due to the resistance of the armature coil is equal and opposite to potential applied to the coil.

Suppose applied potential difference $= V$

Back emf developed $= E$

Resistance of armature circuit $= R$

Current in armature $= I$

then, $$V = E + IR \quad ...(1)$$

or $$I = \frac{V - E}{R} \quad ...(2)$$

Thus, at the start when the speed is low, the back emf. is small so that armature current I is large. At full speed the back emf is large.

Starter

At start, sometimes the armature current is sufficiently large and it can damage the armature windings. To prevent this a variable resistance called started is connected in series with the armature when the speed increases, the starter resistance is reduced.

Efficiency of Motor

The input power used in the following ways,

1. Produces heat in the armature *i.e.*, heat loss I^2R
2. In magnetising the core of the armature is magnetic loss.
3. Mechanical power available.

If V is potential difference and I is current in armature, then,

Input power $= VI$

$$VI = I^2R + \text{magnetic loss} + \text{mechanical power}$$

If magnetic losses are negligible, we have,

$$VI = I^2R + \text{Mechanical power}$$

or mechanical power available,

$$= VI - I^2R$$

$$= I(V - IR), \quad \text{from equation (3)}$$

$$= IE,$$

$$E = V - IR = \text{back emf}$$

Efficiency is defined as the ratio of the output mechanical power to the electrical power

$$= \frac{\text{available mechanical power}}{\text{Input electrical power}}$$

$$= \frac{IE}{IV} = \frac{E}{V}$$

$$= \frac{\text{Back emf}}{\text{Applied emf}} \qquad \text{...(4)}$$

This shows that efficiency increases as back emf E increases. Efficiency is maximum *i.e.*, 100 per cent,

When E = V, but from the relation E = V – IR, under this condition, I = 0, and motor will do no work because the current is zero, hence this cannot be realised in practice.

However, the efficiency is maximum when,

$$EI = \text{maximum}$$

or $$E\left(\frac{V-E}{R}\right) = \text{maximum, (from eq. 2)}$$

or $$\frac{EV}{R} - \frac{E^2}{R} = \text{maximum}$$

Differentiating with respect to E, we get,

$$\frac{V}{R} - \frac{2E}{R} = 0$$

or $$V = 2E$$

or $$E = \frac{V}{2} \qquad \text{...(5)}$$

Hence for maximum efficiency, the back emf should be half the applied voltage. In actual practice this condition cannot be satisfied as the armature resistance being small, the armature current will become high and burn the coil of the armature. Therefore, the back emf in actual practice is allowed to go upto 90 to 95 per cent of the applied emf to keep armature current within the safe limit.

ENERGY DENSITY OF ELECTROMAGNETIC WAVE AND POYNTING THEOREM

One important characteristic of an electromagnetic wave is that it can transport energy from point to point. The direction of electromagnetic wave at a given point is the direction in which energy is being transmitted. We may describe a wave in terms of the new vector $\vec{S}$ which is defined as follows:

$$\vec{S} = \frac{1}{\mu_0}(\vec{E} \times \vec{B}) \text{ or } (\vec{E} \times \vec{H})$$

By this definition the direction of a wave is the direction of the vector $\bar{S}$ *called the "poynting vector"* after John Henry Poynting who first pointed out its properties. Let us now calculate the divergence of Poynting vector $\bar{S}$ for electromagnetic field in free space.

$$\nabla.\frac{1}{\mu_o}(\vec{E}\times\vec{B}) = \frac{1}{\mu_o}(\vec{B}.(\nabla\times\vec{E}) - \vec{E}.(\nabla\times\vec{B})) \text{ [see eq. 3.9]}$$

Substituting $\nabla \times \vec{E}$ and $\nabla \times \vec{B}$ from eq. (1.20), we get,

$$\nabla.\frac{1}{\mu_o}(\vec{E}\times\vec{B}) = \vec{E}\times\vec{B}\nabla.\frac{1}{\mu_o}\left(-\vec{B}.\frac{\delta\vec{B}}{\delta t} - \vec{E}_0.\mu_0\frac{\delta\vec{D}}{\delta t}\right)$$

$$= -\frac{1}{\mu_o}\left[\mu_o\varepsilon_o\bar{E}.\frac{\delta\vec{E}}{\delta t} + \bar{B}.\frac{\delta\vec{B}}{\delta t}\right] \text{ [since } \vec{D} = \varepsilon_o\vec{E}]$$

$$= -\frac{\delta}{\delta t}\left[\frac{1}{2}\varepsilon_o E^2 + \frac{1}{2\mu_o}B^2\right]$$

or $$\nabla.(\bar{E}\times\bar{H}) = -\frac{\delta}{\delta t}\left[\frac{1}{2}\varepsilon_o E^2 + \frac{1}{2}\mu_o H^2\right] \text{ (since B = moH)}$$

On integrating this over the volume v enclosed by the surface S, we get,

$$\int_v \nabla.(\bar{E}\times\bar{H})dv = -\frac{\delta}{\delta t}\int_v\left(\frac{1}{2}\varepsilon_o E^2 + \frac{1}{2}\mu_o H^2\right)dv \qquad ...(1)$$

The left hand side can be transferred to the surface integral by divergence theorem, we get,

$$\int_8 (\bar{E}\times\bar{H}).d\bar{A} = -\frac{\delta}{\delta t}\int_v\left(\frac{1}{2}\varepsilon_o E^2 + \frac{1}{2}\mu_o H^2\right)dv \qquad ...(2)$$

where $d\bar{A}$ is area.

For any other medium of permittivity e and permeability m, we have,

$$\int_8 (\bar{E}\times\bar{H}).d\bar{A} = -\frac{\delta}{\delta t}\int_v\left(\frac{1}{2}\varepsilon E^2 + \frac{1}{2}\mu H^2\right)dv \qquad ...(3)$$

As we know that electrostatic field energy per unit volume is $\frac{1}{2}\varepsilon E^2$ $\left(\text{or } \frac{1}{2}\varepsilon_0 E^2 \text{ for free space}\right)$ and the magnetostatic field energy per unit volume is $\frac{1}{2}\mu H^2$ $\left(\text{or } \frac{1}{2}\mu_0 H^2 \text{ for free space}\right)$. We shall assume that these expressions are valid even though the electric and magnetic fields vary rapidly. Thus the total electromagnetic energy in a given volume in free space is given by,

$$W = \frac{1}{2}\int_v (\varepsilon_0 E^2 + \mu_0 H^2) dv \qquad ...(4)$$

therefore from eq. (1.44a)

$$\int_s (\vec{E} \times \vec{H}) dA = -\frac{dW}{dt}$$

The term – dW/dt represents the rate at which energy is decreasing within the given volume. *The rate of decrease of energy must be equal to the net rate at which energy is flowing out of the volume. Therefore the left hand side represents the total outward flux of energy in watts over the surface enclosing the volume.* The $\vec{S}$ represents the amount of power per unit area. The integral of $\vec{S}$ over the closed surface clearly represents the rate at which energy leaves the enclosed volume.

Thus, for any arbitrary closed surface, the amount of power flowing out through the surface is

$$\oint \vec{S}.d\vec{A}$$

where $d\vec{A}$ is an element of area, therefore,

$$-\frac{dW}{dt} = \oint \vec{S}.d\vec{A} \qquad ...(5)$$

Equation (5) is known as "Poynting's theorem". It states an important property of $\vec{S}$ (the Poynting vector).

$$\vec{S} = \frac{1}{\mu_o}(\vec{E} \times \vec{B}) \text{ or } (\vec{E} \times \vec{H}) \text{ watts/metre}^2$$

represents the power flowing through a unit area normal to the direction of propagation is known as Poynting vector.

Be careful to note that i some situations $\vec{E} \times \vec{H}$ does not represent energy flow, as for example, in a static magnetic field superimposed on a static electric field. Hence *the interpretation that the Poynting vector is representing the flow of energy is not absolute or regorous rule but usually correct.*

ENERGY PER UNIT VOLUME

In Fig 4.7 an area A is normal to the x-axis if a plane wave is travelling in the positive x-direction with speed v, then all the energy contained in the volume A v dt will pass through the area A in time dt.

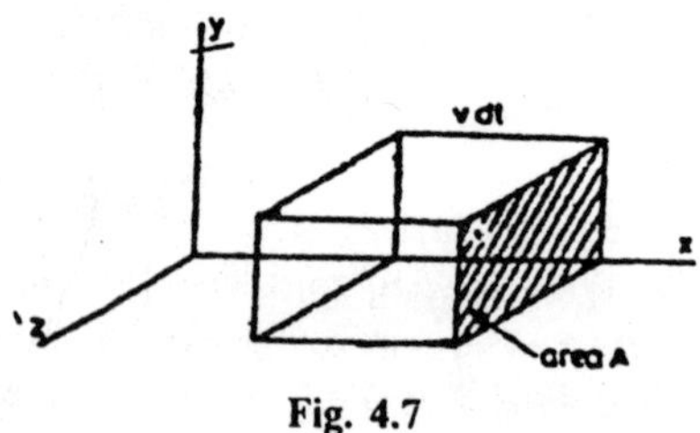

Fig. 4.7

Thus, if U is energy per unit volume, the total energy contained in the volume A v dt is U A v dt. Therefore the energy passing through unit area per unit time is,

$$S = \frac{UA\,v\,dt}{A\,dt} = Uv$$

Thus, $\quad U = S/v \qquad ...(1)$

THE RETARDED POTENTIALS

We have shown that Maxwell's equations have wave type solutions and that the velocity of propagation is C in free space. The plane waves have been treated without inquiring how these waves were produced. We have not shown, however when these solutions are physically relevant or meaningful, nor how they are related to the charges and the currents that must be their sources. Hence our next step is to consider prescribed charge and current distributions, ρ (r, t) and $\vec{J}$ (r, t) and find the fields produced by them that answer above questions. There are several ways of approaching the problem, of which the most fruitful is the potential approach, which is developed analogously to the procedures used in electrostatics and magnetostatics. Therefore if we which to relate the fields of the changing charge and current distribution which produce them it is far easier to work with the potentials than with the fields directly. To be more specific, we shall discuss the generation of electromagnetic waves. We will determine the potentials in terms of given charge and current distribution. Once the potentials have been determined the fields can be obtained from the relation given in Eqs. (1) and (2).

Let the point P on Fig. 4.8, have a position vector r and the point Q have position vector r'. If the charge distribution near Q changes with time, the information that it has changed only be appreciated at the point P after the time r – r'/C. This is the time taken by electromagnetic radiation to travel the distance r – r'.

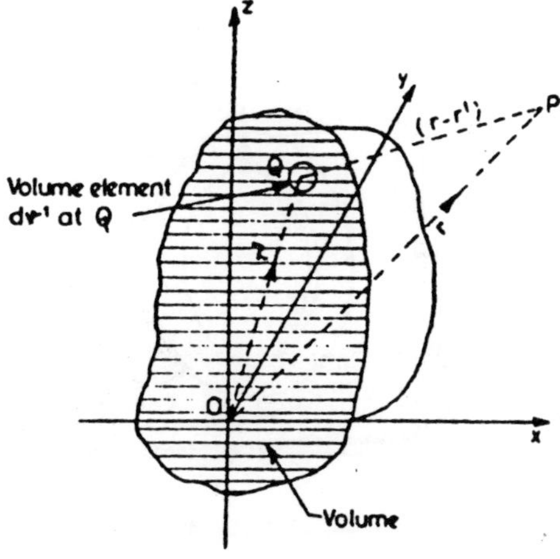

Fig. 4.8

It means that at a time t the contribution to the potential V(r, t) at P from the charge within the volume element dv' at Q depends not on what the charge is at time t but on what it was at the time $t - \dfrac{|r - r'|}{C}$. This is true for all he volume elements within the considered volume. Note that the time lapse $\dfrac{r - r'}{C}$ depends on the position vector r' and so varies over the total volume. The potential is given by

$$V\,(r,\, t) = \frac{1}{4\pi\varepsilon_0} \int_{\text{all}} \frac{\rho\left(r',\, t - \dfrac{|r - r'|}{C}\right)}{|r - r'|} dv' \qquad \text{...(1)}$$

which is known as the "retarded scalar potential".

EARTH INDUCTOR

It is an important device based on electromagnetic induction. It consists of a circular coil C (Fig. 4.9) of 1000 to 5000 turns of insulated copper wire wound found a wooden frame of about 25 cm diameter. The circular frame is capable of rotation about an axis passing through two diametrically opposite points. There is a spring key which allows the coil to rotate through 180° only. The two ends of coil are connected to two terminals T_1 and T_2. They are sometimes also joined to a ballistic galvanometer through a commutator. The handle is used to bring the coil to its position so that on pressing the release key the coil is set in rotation.

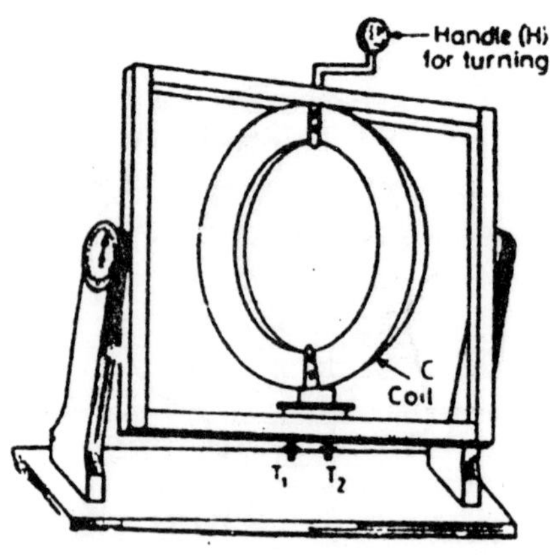

Fig. 4.9

Theory

Consider, a coil of n turns and area A is placed in a magnetic field B. When this coil is rotated in this uniform magnetic field, the magnetic flux associated with it changes and according to Faraday's law an induced emf is set up in it. Suppose at any instant the plane of coil is at right angles to the field, then flux linked with coil = nBA.

Suppose at any instant it makes an angle θ with a direction at right angles to that of the coil (Fig. 4.10) then the flux linked with it nBA cos θ.

Since flux = normal component × area

The induced emf,

$$\xi = -\frac{d\phi}{dt} = -\frac{d}{dt}(nBA\cos\theta)$$

$$= nBA\sin\theta\frac{d\theta}{dt}$$

If R is the resistance of the circuit, induced current is given by,

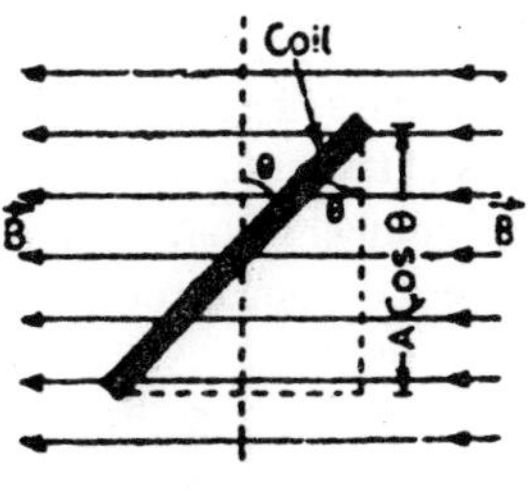

Fig. 4.10

$$I = \frac{\xi}{R} = \frac{nBA}{R}\sin\theta\,\frac{d\theta}{dt}$$

Therefore, the charge that flows through B.G. during a small interval of time dt is,

$$dq = Idt$$

Substituting I,

$$dq = \frac{nBA}{R}\sin\theta\, d\theta$$

when the coil is rotated through half the complete rotation *i.e.*, 0 to 180°, then the total charge passing through the ballistic galvanometer is given by,

$$\int dq = \int_0^{\pi} \frac{nBA}{R}\sin\theta\, d\theta$$

$$q = \frac{nBA}{R}\left[-\cos\theta\right]_0^{\pi}$$

$$= -\frac{nBA}{R}[\cos\pi - \cos 0]$$

$$= \frac{2nBA}{R} \qquad ...(1)$$

If θ_1 is the corresponding first throw in the ballistic galvanometer,

$$\frac{2nBA}{R} = k\theta_1\left(1+\frac{\lambda}{2}\right) \qquad ...(2)$$

where K is the ballistic constant and λ is the logarithmic constant.

STANDARD MAGNETIC FLUX

Standard magnetic fluxes are used for calibrating ballistic galvanometer. The earth inductor also works as a source standard flux is discussed below.

Standard Solenoid or Flux Meter

It consists of along solenoid of large number of turns as primary coil PP. The secondary coil SS of few turns wound over primary coil as shown in Fig. 4.11. Suppose the number of turns per unit length of the primary be n_p. When a steady current I is sent through the primary. The magnetic field the solenoid is given by,

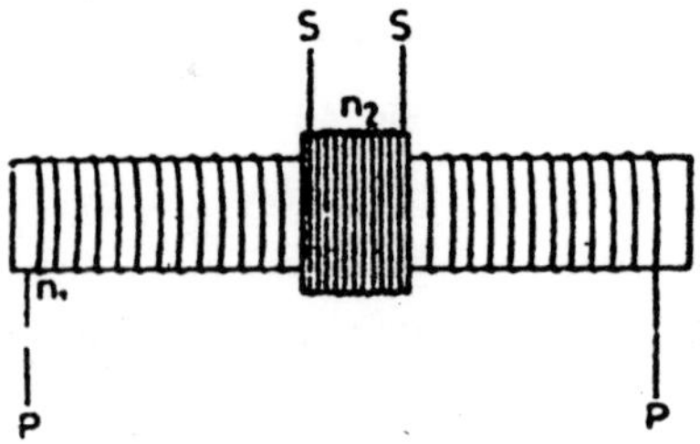

Fig. 4.11

$$B = \mu_o n_P I$$

Now, if A_s be the mean area of the secondary coil and N_s the total number of turns in it. As the lines of magnetic field $\vec{B}$ inside the solenoid also pass through the secondary, producing the magnetic flux through the secondary. The total flux through the secondary is,

$$\phi = BA_sN_s$$

The magnitude of the induced emf

$$\xi = \frac{d\phi}{dt} = \mu_o npN_sA_s\frac{dI}{dt}$$

Induced current $= \dfrac{\xi}{R} = \dfrac{\mu_o npN_sA_s}{R}\dfrac{dI}{dt}$

where R is the total resistance of the secondary,

And the charge q is given by

$$q = \int_0^I \frac{\mu_o npN_sA_s}{R}\frac{dI}{dt}dt$$

$$= \frac{\mu_o npN_sA_sI}{R}$$

If θ is the observed throw in B.G., when this charge passes through it, then

$$\frac{\mu_0 npN_s A_s I}{R} = k\theta_H \left(1+\frac{\lambda}{2}\right) \qquad (1)$$

This equation is used for determining the ballistic constant K. Thus, the galvanometer is calibrated.

USES OF EARTH INDUCTOR

1. Measurement of Horizontal Component H of the Earth Magnetic Field

To measure the horizontal component of the earth's magnetic field at any place, the coil is set with its axis of rotation vertical and the plane of the coil east-west, *i.e.*, perpendicular to the magnetic meridian. The coil is now rapidly rotated through 180° by spring key, consequently induced charge due to change in flux flows through the B.G. which gives a deflection θ_H. From equation (1), we get

$$\frac{2NAH}{R} = k\,\theta_H \left(1+\frac{\lambda}{2}\right) \qquad (1)$$

where R is the total resistance and the coil rotates in field H, *i.e.*, the horizontal component of earth's magnetic field. If the value of K is known, we can determine H. K is calculated by calibrating B.G.

K can be calculated by sending a known current through the primary of standard solenoid and nothing the observed ballistic throw in the galvanometer. The circuit used for this purpose is shown in Fig. 4.12. Let the current in the primary be adjusted to I and the direction of the current is suddenly reversed.

The flux linked with the secondary will change by an amount, given by,

$$\mu_o N_p N_s A_s I - (\mu_o N_p N_s A_s I) = 2\mu_o N_p N_s A_s I$$

charge due to this change in flux,

$$q = \frac{2\mu_o N_p N_s A_s I}{R}$$

If θ is the ballistic throw when this charge passes through ballistic galvanometer,

$$\frac{2\mu_o N_p N_s A_s I}{R} = k\theta\left(1+\frac{\lambda}{2}\right) \qquad ...(2)$$

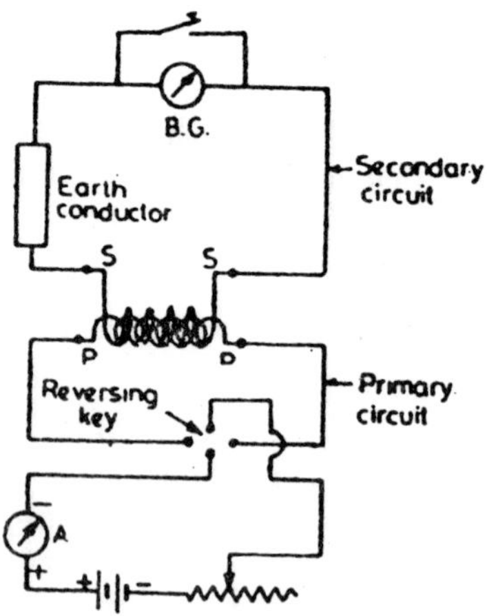

Fig. 4.12

Dividing equation (3.4) by (3.5),

$$\frac{NAH}{\mu_o N_p N_s A_s I} = \frac{\theta_H}{\theta}$$

or

$$H = \frac{\mu_o N_p N_s A_s I}{NA} \frac{\theta_H}{\theta} \text{ weber / metre}^2 \qquad \text{...(3)}$$

Thus, H the horizontal component of the earth magnetic field is known in terms of know quantities.

2. Measurement of Vertical Component V of the Earth Magnetic Field

Here, the coil is adjusted so that is plane is horizontal and it can rotate about a horizontal axis in the magnetic meridian. The coil now rotates in the vertical component V of the earth's magnetic field. The coil is now rapidly rotated through 180° and the deflection θ in the ballistic galvanometer is observed. Proceeding exactly as in previous case, a relation for V is obtained *i.e.*,

$$V = \frac{\mu_o N_p N_s A_s I}{NA} \frac{\theta_v}{\theta} \text{ weber / metre}^2 \qquad \text{...(4)}$$

The intensity of the earth's magnetic field

$$= \sqrt{H^2 + V^2}$$

$$= \frac{\mu_o N_p N_s A_s I}{NA} \frac{1}{\theta} \sqrt{\theta_H^2 + \theta_V^2} \qquad ...(5)$$

3. Measurement of the Angle of Dip

The above two experiments are performed and the angle of dip is given by

$$\tan \delta = \frac{V}{H}$$

Dividing equation (2) by (3), we get,

$$\tan \delta = \frac{V}{H} = \frac{\theta_V}{\theta_H} \qquad ...(6)$$

The calibration of galvanometer is not needed in this case.

4. Calibration of Ballistic Galvanometer Using Earth Inductor

Equations (2) can be used for the calibration of the ballistic galvanometer, because if H is known we can determine the constant K as given below :

$$\frac{2NAH}{R} = k\, \theta_H \left(1 + \frac{\lambda}{2}\right)$$

$$K = \frac{2NAH}{R\theta_H \left(1 + \frac{\lambda}{2}\right)} \qquad ...(7)$$

MEASUREMENT OF A STRONG MAGNETIC FIELD USING A SEARCH COIL

A search coil is a solenoid of small diameter having few (about 50) turns of insulated copper wire and is suitable for measuring strong magnetic field.

Suppose we have to find the magnetic field between the pole-pieces of an electromagnet. The search coil is placed in the gap between the pole pieces and is connected in series with a ballistic galvanometer through a high resistance box R to control the throw in B.G. and the secondary coil SS of standard solenoid as shown in Fig. 4.13. We can change the magnetic flux in two ways.

(1) Put on the magnetic field of the electromagnet and the search coil is either suddenly inserted in or withdrawn from the field, *i.e.*, the region between pole pieces NS, with its plane at right angles to the magnetic field.

(2) The search coil is held between the pole pieces and the magnetic field is suddenly put on.

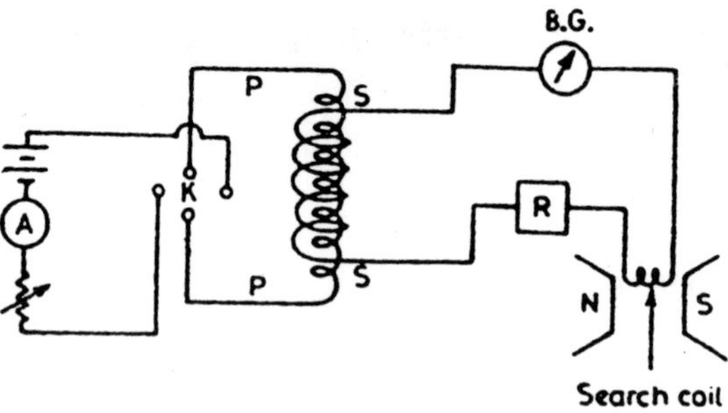

Fig. 4.13

In both cases the flux linked with the search coil changes and produces an induced emf in the circuit which gives throw in ballistic galvanometer.

Suppose the throw	$= \theta$
Number of turns of the search coil	$= N$
The area of each turn of the coil	$= A$
The magnetic field (or flux density)	$= B$
Then flux linked with the search coil	$= NAB$

But this represents change in flux in both cases mentioned above, since in the first case initially the flux was zero whether the coil was outside or inside, as the case may be. In the second case mentioned above the flux becomes zero when the field is put off. Thus in both cases,

The change in flux $d\phi = NAB$

Induced emf in time $dt = \dfrac{NAB}{dt}$

$$\text{Current} \quad I = \frac{NAB}{Rdt}$$

The charge flowing in time dt

$$q = Idt = \frac{NAB}{R}$$

This charge flows through B.G. them,

$$\frac{NAB}{R} = K\theta\left(1 + \frac{\lambda}{2}\right) \qquad ...(1)$$

In order to determine k, *i.e.*, for calibrating the galvanometer a series circuit containing a battery, an ammeter and a rheostat is connected to the primary of the standard solenoid. A known current in the primary circuit is passed. It produces the field inside the solenoid $\mu_o n_p I$. This causes a magnetic flux to pass through the secondary. The flux linked with secondary coil = $\mu_o n_p N_s A_s I$.

where n_p are the number of turns per unit length in the primary coil,

N_s the total number of turns in the secondary coil

A_s the area of each secondary coil and I is the current through the primary.

Now, the direction of current in the primary is suddenly reversed with reversing key, the total change in flux,

$$= \mu_o n_p N_s A_s I - (\square\, \mu_o n_p N_s A_s I)$$

$$= 2\, \mu_o n_p N_s A_s I$$

As above, the charge due to this change in flux

$$= \frac{2\, \mu_o\ n_p\ N_s\ A_s\ I}{R}$$

Let θ_1 be the throw observed in ballistic galvanometer when the charge passes through it then,

$$= \frac{2\, \mu_o\ n_p\ N_s\ A_s\ I}{R} = k\,\theta_1 \left(1 + \frac{\lambda}{2}\right) \qquad ...(2)$$

Dividing eqn. (1) by (2), we get,

$$= \frac{NAB}{2\, \mu_o\ n_p\ N_s\ A_s\ I} = \frac{\theta}{\theta_1}$$

$$\text{or} \quad B = \frac{2\, \mu_o\ n_p\ N_s\ A_s\ I}{NA} = \frac{\theta}{\theta_1}\ \text{weber / metre}^2 \qquad ...(3)$$

Hence, B can be determined.

While calibrating the galvanometer, if instead of reversing the direction of current in the primary, the current is suddenly started or stopped, the change in flux linked with secondary will be $\mu_o n_p N_s A_s I$ and the throw θ_2 in B.G. will now be half of the θ_1, thus,

$$B = \frac{2\, \mu_o\ n_p\ N_s\ A_s\ I}{NA} = \frac{\theta}{\theta_2} \qquad ...(4)$$

Hence, B can be determined.

BASIC EQUATIONS

The basic equations of electricity and magnetism which we have studied are summarised in differential form by the following four equations.

(1) Gauss's law for electric field of charge, *i.e.*,

$$\text{div } \vec{D} = \nabla . \vec{D} = r \quad ...(1)$$

(2) Gauss's law for magnetic field *i.e.*,

$$\text{div } \vec{B} = \nabla . \vec{B} = 0, \quad ...(2)$$

(3) Faraday's law for induced emf produced by the rate of change of magnetic flux, *i.e.*,

$$\text{Curl } \vec{E} = \nabla \times \vec{E} = -\frac{\delta \vec{B}}{\delta t}, \quad ...(3)$$

(4) Ampere's law for magnetic field due to Transformers and A.C. Bridges *i.e.*,

$$\text{Curl } \vec{B} = \nabla \times \vec{B} = \mu_o \vec{j} \quad ...(4)$$

The first three equations are valid for static as well as dynamic field and they apply in general. However, the fourth law *i.e., Ampere's law was derived for Transformers and A.C. Bridges and does not hold for time varying field.*

TYPES OF CURRENTS

We shall the discuss currents in material in a manner similar to that used to treat charges in material medium. In general, currents may be classified in two categories:

(1) *True currents* due to the motion of true charges.

(2) *Other currents* which are due to medium itself.

This separation is analogous to the separation that was discussed in the electrostatic theory between the potential (or fields) of true charges and the potentials of polarisation charges. Similarly here, there are two types of magnetic fields, one derived from the true currents and the other derived from the combined effects of all the currents whatever may be their origin. It is this latter field, namely, the field of magnetic induction $\vec{B}$, that can be considered to be space-time average of the interatomic fields.

All currents are classified as follows:

(1) *True currents* indentical with the physical transport of true charges.

(2) *Polarisation currents.* Currents those arise from the change of polarisation with time.

(3) *Magnetisation currents.* These are stationary currents that flow within regions that are inaccessible to observation but might give rise to net boundary or volume currents, due to imperfect orbit cancellation on an atomic scale.

(4) *Convective currents.* If a material medium is in motion and contains charges of various types. Additional currents are obtained those arise from the motion of both true and polarisation charges contained in the medium. Such currents are called convective currents.

Therefore in a stationary medium the total current is given by the sum of the first three types of currents given above. Thus,

$$\vec{J}_{total} = \vec{J}_{true} + \frac{\delta \vec{P}}{\delta t} + \nabla \times \vec{M} \quad ...(1)$$

VACUUM DISPLACEMENT CURRENT

Taking the divergence of equation (1), we get,

$$\text{div curl } \vec{B} = \mu_o \text{ div } \vec{J}$$

The left hand side will become zero. *Since the divergence of curl of a vector is always zero.* Hence, the right hand side is also zero,

$$\text{div } \vec{J} = \nabla.\vec{J} = 0 \quad ...(1)$$

Here $\vec{J}$ is total current as given in equation (1).

This means that the total flux of current out of nay closed surface is zero. Physically, this means that current is always closed and there are no sources or sinks. Thus we arrive at a contradiction. Since for time varying case *i.e.,* for changing electric fields such as the discharge of a charged condenser through a conducting wire joining its plates. The current starts at positively charged plate, whose charge gradually decreases as the current flows through the conducting wire to the negatively charged plate and neutralises the negative charge on it. Thus the condenser plates are regarded as sources or sinks. *Hence the Ampere's law is valid only for steady state phenomena and not for changing fields.*

Equation (2) is also in conflict with the equation of continuity which gives the complete relation

$$\text{div}\ \vec{J} = \frac{\delta\rho}{\delta t} \qquad ...(2)$$

Maxwell concluded that equation (1) is incomplete. The contradiction of eq. (1) with the equation of continuity *i.e.,* (2) could be removed by converting the equation of continuity into a form whose divergence is zero by adding something to $\vec{j}$ in equation (2). Hence to assume the conservation of charge, it is necessary that total current should obey the equation of continuity

Gauss's law is

$$\nabla.\vec{D} = \rho$$

Differentiating with respect to time

$$\frac{\delta}{\delta t}\nabla.\vec{D} = \frac{\delta\rho}{\delta t}$$

$$\nabla.\frac{\delta\vec{D}}{\delta t} = \frac{\delta\rho}{\delta t}$$

Adding $\nabla.\vec{J}$ on both sides,

$$\nabla.\vec{J} + \frac{\delta\rho}{\delta t} = \nabla.\vec{J} + \nabla.\frac{\delta\vec{D}}{\delta t}$$

$$= \nabla.\left(\vec{J} + \frac{\delta\vec{D}}{\delta t}\right)$$

where $\vec{D} = \varepsilon_0\vec{E}$ is displacement vector in free space. According to equation of continuity,

$$\nabla.\vec{J} + \frac{\delta\rho}{\delta t} = \nabla.\left(\vec{J} + \frac{\delta\vec{D}}{\delta t}\right) = 0$$

Thus $\nabla.\vec{J} = 0$, eq. (1.6) for Transformers and A.C. Bridges and $\nabla.\left(\vec{J}+\frac{\delta\vec{D}}{\delta t}\right) = 0$, every where ...(3)

i.e., for steady currents as well as currents corresponding to time varying fields.

So that Maxwell replaced $\vec{J}$ in Ampere's law by,

$\vec{J}+\frac{\delta\vec{D}}{\delta t}$ and Ampere's law becomes

$$\text{curl } \vec{B} = \mu_o\left(\vec{J}+\frac{\delta\vec{D}}{\delta t}\right) \quad ...(4)$$

The term $\frac{\delta\vec{D}}{\delta t}$ is called is the *displacement current density* since it arises when the electric displacement $\vec{D}$ (*i.e.*, in effect $\vec{E}$) changes with time. Maxwell assumed that it is just as effective as $\vec{J}$ in producing magnetic field. The introduction of displacement current is one of Maxwell major contribution to the electromagnetic theory without which the, electromagnetic theory would have been impossible.

In fact, relation (4) is called Maxwell's equation as derived from Ampere's law.

Significance of Displacement Current

The vacuum displacement current $\varepsilon_o \frac{\delta E}{\delta t}$ does not have the significance of a current in the sense of being the motion of charges, with the introduction of displacement of current the total current given in equation (5) is,

$$\vec{C} = \vec{J}_{true} + \frac{\delta\vec{P}}{\delta t} + \nabla\times\vec{M} + \varepsilon_o\frac{\delta\vec{E}}{\delta t}$$

$$\vec{C} = \vec{J}_{total} + \varepsilon_o\frac{\delta\vec{E}}{\delta t} \quad ...(5)$$

The current C obtained by adding the term $\varepsilon_o \frac{\delta\vec{E}}{\delta t}$ to the total current, is a solenoidal current, of which divergence is zero. The introduction of displacement current is necessary in order to apply the Ampere's law for magnetic field due to non-stationary currents.

The geometrical significance of the solenoidal current C is that at points where there is an accumulation of charge, the current is assumed to be continuous across the discontinuity in the form of the rate of change of the field resulting from the accumulation of charge on the boundaries of the discontinuity. As an example charging a capacitor produces a closed current loop in terms of C.

Examples of Displacement

The idea of displacement current originated from the study of discharge of condenser. Before Maxwell the discharge current was supposed to flow in the open circuit formed by the condenser plates and the wires. Maxwell interpreted that all electric currents are divergenceless *i.e.*, electric currents must be closed and in the discharge of condenser the conduction current is closed by displacement current (arising due to change of electric displacement in the dielectric) flowing in the dielectric between the plates.

Consider a voltage applied to a resistor and a capacitor in parallel as shown in Fig. 4.14.

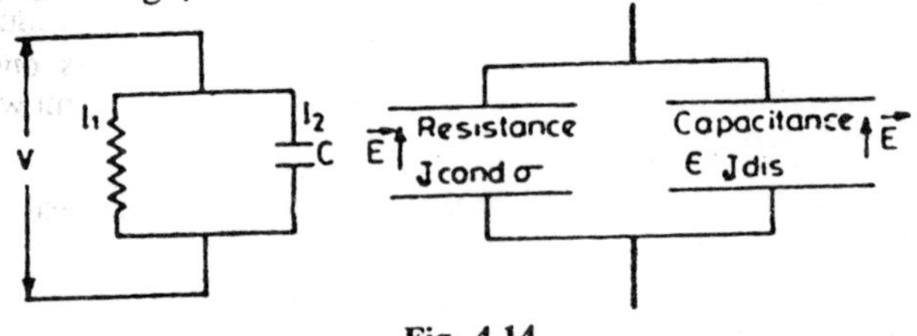

Fig. 4.14

The nature of current flowing through the resistor is different from that through the capacitor. A constant voltage across a resistor produces a continuous flow of current of constant value, on the other hand, the current through a capacitor will be constant only when the voltage is changing.

The current I_1 through the resistor is given by,

$$I_1 = \frac{V}{R}$$

and the current through the capacitor is given by,

$$I_2 = \frac{dq}{dt} = C\frac{dV}{dt}$$

Since the instantaneous charge in the capacitor is given by $q = CV$.

The current through the resistor is a conduction current, while the current through the capacitor is displacement current. Although, the current does not flow through the capacitor, the external effect is as though it did. Since as much current flows out of one plate as flows into the opposite one.

The electric field $\vec{E}$ equals the voltage V across the element divided by its length d, that is

$$E = V/d$$

The current density J_1 inside the resistor equals the product of the electric field E and the conductivity of the medium inside the resistor element and also I_1 divided by the cross-sectional area A or

$$J_1 = \sigma E = \frac{I_1}{A}$$

The capacitance of parallel plate capacitor

$$C = \frac{\varepsilon A}{d}, \text{ where } \varepsilon \text{ is permittivity of the medium, then}$$

$$I_2 = \frac{\varepsilon A}{d} \cdot \frac{d}{dt}(Ed) = \varepsilon A \frac{dE}{dt}$$

and current density

$$J_2 = \frac{I_2}{A} = \frac{\epsilon\, dE}{dt}$$

Recalling that $D = \varepsilon E$

$$J_2 = \frac{dD}{dt}$$

In this example J_1 is a conduction current density J_{cond}, while J_2 is a displacement current density J_{disp}. Since the current density $\vec{J}$, the electric displacement $\vec{D}$ and the electric field intensity $\vec{E}$ are actually space vectors which all have the same direction in isotropic media, hence we can write,

$$\vec{J}_{cond} = \sigma \vec{E}$$

$$\text{and } \vec{J}_{disp} = \frac{\varepsilon d\vec{E}}{dt} = \frac{d\vec{D}}{dt}$$

The total current density $\vec{J}_{total}$

$$\vec{J}_{total} = \vec{J}_{cond} + \vec{J}_{disp} \qquad ...(6)$$

Thus there is a displacement current density in the region of condenser plates, *the concept of displacement current has very important consequences for dielectrics and for free space.* The displacement current is responsible for the production of magnetic fields in empty space since the conduction current is zero and the magnetic fields are entirely due to displacement currents. The displacement current cannot be detected directly but the confirmation of its existence is provided indirectly by the electromagnetic theory.

CONDITION WHEN A MOVING COIL GALVANOMETER IS DEAD BEAT OR BALLISTIC

In deriving equation (1) we have assumed that whole of the kinetic energy of the coil is used up in twisting the fibre and neglected the retarding forces present namely, (1) due to friction of air, (2) viscosity of suspension fibre and (3) induced current in any neighbouring mass or metal. The first two of these factors constitute mechanical damping, while the third factor accounts for electromagnetic damping. The retarding couple due to damping varies as the angular velocity = $d\theta/dt$, where p is coefficient of mechanical damping.

When the circuit is closed, we have in addition the induced current produced in the coil which varies as the rate of change of displacement and inversely as the total resistance of the circuit, hence the retarding couple is $\frac{m}{R}\frac{d\theta}{dt}$, where the constant m involves the magnetic flux due to the magnet and the area of coil. Thus, the total retarding couple due to damping

$$= \left(p + \frac{m}{R}\right)\frac{d\theta}{dt}.$$

Now the angular acceleration is $\frac{d^2\theta}{dt^2}$.

Hence, according to Newton's second law, equation of motion is,

$$\frac{Id^2\theta}{dt^2} = -c\theta - \left(p + \frac{m}{R}\right)\frac{d\theta}{dt}$$

or
$$\frac{Id^2\theta}{dt^2} + \left(\frac{m}{R} + p\right)\frac{d\theta}{dt} + c\theta = 0$$

or $$\frac{d^2\theta}{dt^2}+\frac{1}{I}\left(\frac{m}{R}+p\right)\frac{d\theta}{dt}+\frac{c\theta}{I}=0$$

or $$\frac{d^2\theta}{dt^2}+2b\,\frac{d\theta}{dt}+K^2\theta=0 \qquad ...(1)$$

where, $$2b=\frac{1}{I}\left(\frac{m}{R}+p\right) \qquad ...(2)$$

and $$K^2=\frac{c}{I} \qquad ...(3)$$

Let the solution of this equation be $\theta = Ae^{\alpha t}$

Then, $$\frac{d\theta}{dt}=A\alpha e^{\alpha t} \text{ and } \frac{d^2\theta}{dt^2}=A\alpha^2 e^{\alpha t}.$$

Putting these values in equation (1),

$$\alpha^2 + 2b\alpha + K^2 = 0$$

or $$\alpha = -b \pm \sqrt{b^2 - K^2}$$

This gives,

$$\theta = A_1 e^{(-b+\sqrt{b^2-K^2})t} + A_2 e^{(-b-\sqrt{b^2-K^2})t}$$

where, A_2 and A_1 are constants.

The quantity $b^2 - K^2$ may be positive, zero or negative depending upon whether $b > K$, $b = K$ or $b < K$.

Case 1: When $b > K$, we have two real values of a, then solution of equation (1) is,

$$\theta = A_1 e^{\alpha 1t} + A_2 e^{\alpha 2t}$$

The α_1 and α_2 are negative. The deflection θ will go no decreasing with increasing time. *The motion is non-oscillating or dead beat and the oscillation is said to be over-damped.*

Case 2 : Critical damping : When $b = K$, the two values of a are equal to $-b$ and solution of (1) becomes

$$\theta = (A_1 + A_2)\, e^{-bt}$$
$$= A_3\, e^{-bt}$$

That the coil after deflection comes to rest in minimum possible time and the galvanometer is said to be *critically damped. However, the motion of the coil is just non-oscillatory*

Case 3 : When b > K, the quantity $\sqrt{b^2 - K^2}$ is imaginary so the values of a are complex *i.e.*,

$$\alpha = -b \pm jg. \text{ where } g^2 = K^2 - b^2$$

Hence, solution of equation (1) becomes,

$$\theta = A_1 e^{(-b+jg)t} + A_2 e^{(-b-jg)t}$$

$$= e^{-bt}[A_1 e^{jgt} + A_2 e^{-jgt}]$$

$$= e^{-bt}[A_1\{\cos gt + j\sin gt\} + A_2\{\cos gt - j\sin gt\}]$$

$$= e^{-bt}[(A_1 + A_2)\cos gt + j(A_1 - A_2)\sin gt]$$

Putting $A_1 + A_2 = B\sin\beta$

and $j(A_1 - A_2) = B\cos\beta$

$\theta = Be^{-bt}\sin(gt + \beta) = Be^{-bt}\sin\{(\sqrt{K^2 - b^2})t + \beta\}$.

The motion is oscillatory or ballistic and period of oscillation,

$$T = \frac{2\pi}{g} = \frac{2\pi}{\sqrt{K^2 - b^2}} \qquad ...(4)$$

In order that the whole charge may pass through the ballistic galvanometer before it has moved from its zero position, it is essential that its periodic time be large. Since we have,

$$T = 2\pi\sqrt{\frac{I}{c}}.$$

therefore, I should be large and c should be small. As we have

$$b < K$$

$$\frac{1}{2I}\left(\frac{m}{R} + p\right) < \frac{c}{I} < \frac{2\pi}{T}$$

Hence to make b small,

I should be large,

R should be large,

m should be small and so the coil should be wound round a nonconducting frame, *e.g.*, paper, bamboo or plastic. p should be small, and so the air resistance should be small and the fibre should be very fine.

We just see that the requirements of dead-beat galvanometer are just opposite to those for the ballistic b > K. Hence the opposite conditions should prevail. The coil is made of thick copper wire and has smaller number of turns so to have smaller resistance and is wound round a metallic or conducting frame.

SOLVED EXAMPLES

Example 1:

A copper strip 2.0 cm wide and 1.0 mm thick is placed in a magnetic field of magnitude 1.5 webers/ metre^2 *as z-direction. If a current of 100 amp is set up in the strip along the x-direction, calculate the Hall potential difference across the strip given the number of electrons per cubic metre =* 8.4×10^{28} *metre*^{3}.

Solution:

From equation (1),

$$E_y = \frac{j_x B_z}{Ne}$$

But $E_y = \frac{V}{d}$ where V is the potential difference

and $jx = \frac{i}{A} = \frac{i}{dh}$

where h is thickness.

Substituting these values, we get

$$\frac{V}{d} = \frac{i}{dh} \cdot \frac{B_z}{Ne}$$

$$V = \frac{iB_z}{Neh} = \frac{200 \text{ amp} \times 1.5 \text{ weber / metre}^2}{8.4 \times 10^{28/m^3} \times 1.6 \times 10^{-19} \text{ coul} \times 1.0 \times 10^3 \text{ m}}$$

$$= 2.2 \times 10^{-5} \text{ volt}$$

$$= 22 \text{ micro volt.}$$

Example 2:

A uniform magnetic field of magnitude 1.5 Weber/metre^2 *points horizontally from south to north. A proton of energy 5.0 MeV moves vertically downward through this field. Calculate the force on it.*

Solution:

Kinetic energy of the proton is

$$K = 5.0 \text{ MeV} = 5 \times 10^{6} \text{ eV}$$
$$= 5 \times 10^{6} \text{ eV} \times 1.6 \times 10^{-19} \text{ joule}$$
$$= 8.0 \times 10^{-13} \text{ joule.}$$

But K = 1/2 mv^2, where m is the mass and v is the velocity, or

$$v = \sqrt{\frac{2K}{m}} = \sqrt{\frac{2\times8.0\times10^{-13}}{1.7\times10^{-27}}} = 3.1\times10^{7} \text{ metres/sec.}$$

Now, according to equation (1.2),

$$F = qvB \sin \theta.$$

As given $\vec{B}$ is horizontal and proton moves vertically downward, hence angle between $\vec{v}$ and $\vec{B}$ is 90°,

therefore, $\sin \theta = \sin 90 = 1$

F = qvB = (1.6 × 10^{-19}Coul) × (3.1 × 10^{7}metre/sec) × 1.5weber/ metre^2

But, $\text{Weber} = \frac{\text{Newton.metre}}{\text{Ampere}}$

hence, F = 7.4 × 0^{-12} Netwons.

Since $\vec{F}$ is perpendicular to $\vec{v}$ and $\vec{B}$ from relation,

$$\vec{F} = q\vec{v} \times \vec{B},$$

the direction of this force is east.

Example 3(a):

A 10 cm long wire carrying a current of 10 amp. is held at an angle 30° with the direction of a uniform magnetic field of strength 1 weber/ metre². Calculate the force acting on the wire.

Solution:

According to equation (1.6), the force is

$$\vec{F} = i\vec{l} \times \vec{B}$$
$$\vec{F} = ilB \sin \theta.$$

Putting the given values,

$$F = 10 \times 10 \times 10^{-2} \times 1 \sin 30$$

$$= 0.5 \text{ Newton.}$$

Example 3(b):

If the magnitude of $\vec{H}$ in a plane wave is 1 amp./metre. Find the magnitude of $\vec{E}$ for a plane wave in free space.

Solution:

By equation (i)

$$E = \sqrt{\frac{\mu_0}{\varepsilon_0}} H$$

$$= 376 \times 1 \text{ ohms. Amp/metre}$$

$$= 376 \text{ volts/metre.}$$

Example 4:

When 0.1 coul. of charge is passed through a moving coil B.G. a deflection of 30 mm is observed on a scale one meter away. Find the current sensitivity of the galvanometer if the time period of the coil is 10 seconds.

Solution:

From equation (1)

$$\text{Current sensitivity} = \frac{T}{2\pi} \times \text{charge sensitivity}$$

$$= \frac{10 \text{sec}}{2 \times 3.14} \times \frac{30 \text{ mm}}{0.1 \text{ Coul}}$$

$$= 477.5 \text{ mm/amp.}$$

Example 5(a):

A condenser of capacity 0.2 μF is charged to 3 volts when discharged through a B.G., it gives a deflection of 10 cm. Calculate the current sensitivity of galvanometer, if its periodic time be 10 sec.

Solution:

$$\text{Charge sensitivity} = \frac{10}{0.2 \times 3} \frac{\text{cm}}{\text{micro coulomb}}$$

$$= 16.6 \frac{\text{cm}}{\text{micro coulomb}}$$

$$\text{Current sensitivity} = \frac{2\pi}{T}.\ \text{Charge sensitivity}$$

$$= \frac{2 \times 3.14}{10 \text{ sec}} \times \frac{16.6 \text{ cm}}{\text{micro coul}}$$

$$= 10.4 \text{ cm/micro amp.}$$

Example 5(b):

A circular coil of 100 turns has an effective radius 50 cm and carries a current 0.10 amp. Calculate the work done required to turn it in an external uniform magnetic field 1.5 weber/metre^2 through 180°.

Solution:

Let us take the coil be turned from position $\theta = 0$ to $\theta = 180°$, the work required is the difference in energy between the two positions, the potential energy $\vec{U} = -\text{m}.\vec{B}$

$$\text{Hence, work required} = U_e = 180 - U\theta = 0$$

$$= -\text{mBCos } 180 - (-\text{ mB Cos } 0)$$

$$= 2\text{mB}$$

But $\text{m} = \text{NiA}$

$$\text{Hence, work required} = 2\ \text{NiAB}$$

$$= 2\text{Ni}\ (\pi r^2)\ \text{B}$$

$$= 2 \times 100 \times (0.10) \times \pi \times (5 \times 10^{-2})\ 1.5$$

$$= 0.24 \text{ joule}$$

Example 6:

A condenser charged to 2 volts is discharged through a ballistic galvanometer, when the corrected deflection is 1.6 cms and the current sensitivity is 4.54 × 10^2 mm/μA and the periodic time is 12 seconds. Calculate the capacity of the condenser.

Solution:

$$\text{Charge sensitivity} = \frac{T}{2\pi} \times \text{current sensitivity}$$

$$= \frac{12}{2 \times 3.14} \times 4.54 \times 10^2 \text{ mm/micro coul}$$

$$= \frac{12 \times 4.5 \times 10}{2 \times 3.14} \text{ cm/micro coul}$$

But $\quad \text{charge sensitivity} = \frac{\text{deflection}}{\text{charge}}$

or $\quad \text{Charge} = \frac{\text{deflection}}{\text{charge sensitivity}}$

$$= \frac{9.6 \text{ cm}}{\frac{12 \times 4.5 \times 10}{2 \times 3.14} \text{ cm/micro coul}}$$

or $\quad q = \frac{9.6 \times 2 \times 3.14}{12 \times 4.5 \times 10} \; \mu \text{ coul}$

$$= \frac{9.6 \times 2 \times 3.14 \times 10^{-6}}{12 \times 4.5 \times 10} \text{ coul}$$

But, $\quad q = c\,V$

or $\quad c = \frac{q}{V} = \frac{9.6 \times 2 \times 3.14 \times 10^{-6}}{12 \times 4.5 \times 10 \times 2} \text{ Coul/volt}$

$$= 20.16 \times 10^{-8} \text{ Farad}$$

$$= 0.2016 \; \mu\text{F}.$$

Example 7:

The successive deflections to the right and lift of the mean position in the case of ballistic galvanometer are found to be 25.9 and 24.8 cms. Calculate the logarithmic decrement.

Solution:

As we know that $\quad \frac{\theta_1}{\theta_2} = \frac{\theta_2}{\theta_3} e^{\lambda}$

$$e^{\lambda} = \frac{25}{24.9} = \frac{24.9}{24.8} = \frac{25 + 24.9}{24.9 + 24.8}$$

$$\lambda = 2.3 \log_{10} \frac{49.9}{49.7} = 0.0039$$

Example 8:

A condenser of 1,000 of is charged to a potential difference of 1 volt and then discharged through a B.G. The first throw an noted on a scale placed 1 meter away is 62.2 cms. If the time period of free vibrations be 10 seconds, and logarithmic decrement 0.02, calculate the ballistic constant of the galvanometer and its figure of merit.

Solution:

$$q = cV$$

$$= 1000 \times 10^{-12} \times 1 = 10^{-9} \text{ coul.}$$

and also
$$q = K\theta_1 \left(1+\frac{\lambda}{2}\right),$$

where,

$$K = \frac{T}{2\pi} \frac{c}{NAB}$$

or,

$$10^{-9} = K \times 62.2 \times 10^{-2} \times \left(1+\frac{0.02}{2}\right)$$

or
$$K = \frac{10^{-9}}{62.2 \times 10^{-2} \times \left(\frac{2+0.02}{2}\right)}$$

$$= \frac{10^{-9} \times 1000 \times 100}{622 \times 101} \frac{\text{Coul}}{\text{metre}}$$

Since,
$$K = \frac{T}{2\pi}\left(\frac{c}{NAB}\right)$$

In fact, $\frac{c}{NAB}$ is the figure of merit

$$\frac{c}{NAB} = K.\frac{2\pi}{T} = \frac{10^{-9} \times 1000 \times 100}{622 \times 101} \frac{2\pi}{10}$$

$$= 10^{-10} \text{ amp/radian},$$

Let us consider a deflection of 1 mm on a scale, 100 cm away

$$\text{Angular deflection} = \frac{10.3}{1} = \frac{1}{1000} \text{ radian}$$

Current required to produce this deflection,

$$= 10^{-10} \text{ amp/radian} \times \frac{1}{1000} \text{ radian}$$

$$= 10^{-13} \text{ amp.}$$

Example 9:

A 50 ohm coil galvanometer can carry a current 10 MA. How can you convert it into

(a) voltmeter of range 500 volts

(b) ammeter of range 10 amp.

Solution:

(a) Using relation (1.31), *i.e.*,

$$V = i_g (R_m + R_g) \text{ or } R_m = \frac{V}{i_g} - R_g$$

$$i_g = 10\text{MA}$$

$$= 10 \times 10\text{–}3 \text{ amp}$$

$$R_g = 50 \text{ ohms}$$

$$V = 500 \text{ volts}$$

hence,

$$R_m = \frac{500}{10\times10^{-3}} - 50$$

$$= 49{,}950 \text{ ohms}$$

R_m = 49950, is connected in series.

(b) Using relation (1.30), *i.e.*,

$$i = i_g\left(1+\frac{R_g}{R_s}\right)$$

given,

$$i = 10 \text{ amp}$$

$$i_g = 10 \times 10^{-3} \text{ amp.}$$

$$R_g = 50 \text{ ohms}$$

Hence,

$$10 = 10 \times 10^{-3}\left(1+\frac{50}{R_s}\right)$$

or

$$\frac{10}{10\times10^{-3}} = \frac{R_s+50}{R_s}$$

or

$$1+\frac{50}{R_s} = 1000$$

or

$$\frac{50}{R_s} = 999$$

or $$R_s = \frac{50}{999} = 0.05005$$

R_s is shunt resistance.

Example 10:

A pointer type galvanometer of 20 ohm resistance has a linear scale of 50 divisions. It gives a deflection of 30 divisions when connected in series with a resistance of 2980 ohms and an accumulator of 2 volts. How will you convert it into an ammeter of range 3 ampere ?

Solution:

The current corresponding to 30 divisions by Ohm's law.

$$i_{30} = \frac{2}{2980+20} = \frac{2}{3000} \text{ amp.}$$

The maximum value of the current galvanometer can carry is corresponding to 50 divisions. Hence,

$$i_g = i_{50} = \frac{2}{3000} \times \frac{50}{30} \text{ amp.}$$

Now using the relation,

$$i = i_g\left(1+\frac{R_g}{R_s}\right)$$

$$3 = \frac{2}{3000} \times \frac{50}{30}\left(1+\frac{20}{R_s}\right)$$

or $$\frac{3\times3000\times30}{2\times50} = 1+\frac{20}{R_s}$$

or $$\frac{20}{R_s} = \frac{3\times3000\times30}{2\times50} - 1$$

$$= 2700 - 1 = 2699$$

or $$R_s = \frac{29}{2699}\ 0.00741 \text{ ohm (shunt)}$$

Example 11:

Paraffin has relative permittivity K = 2.1. Find the index of refraction for paraffin and also the velocity of the wave in paraffin.

Solution:

As paraffin is a non-ferrous material, hence by eq. (1)

$$\eta = \sqrt{K} = \sqrt{2.1} = 1.45$$

By Eq. (1.30) $v = C/\eta = C/\sqrt{K}$

$$= C/\sqrt{2.1} = 2.07 \times 10^6 \text{ metre/sec.}$$

Example 12:

The relative permittivity of distilled water is 81. Calculate refractive index and velocity.

Solution:

$$\eta = \sqrt{K} = \sqrt{81} = 9$$

$$v = C/\sqrt{81} = 0.11\ C$$

$$= 3.33 \times 107 \text{ metres/sec.}$$

EXERCISES

1. State Maxwell's equations and solve them to obtain the velocity of the electromagnetic waves in a homogeneous isotropic dielectric medium.
2. Deduce theoretically Poynting theorem for the flow of energy in an electromagnetic field,
3. Write the Maxwell's equations. Explain the physical significance of each equation.
4. Explain different types of currents you know. Explain fully the concept of displacement current.
5. Write Maxwell's equations. Solve them in free space.
6. Describe the principle, construction and working of an A.C. generator or dynamo.
7. What do you mean by a D.C. generator? Describe its working.
8. What are energy losses in a generator? How these are reduced?
9. Describe the principal and construction of an D.C. electric motor. Define the efficiency of an electric motor. Prove that for maximum output, the back emf should be half the applied voltage.

10. Describe with necessary theory the use of an earth inductor to find the angle of dip.
11. Describe with relevant theory the method for measuring strong magnetic field.
12. Discuss the theory and construction of a Grassot's fluxmeter. How it is used in actual practice?
13. Deduce Maxwell's equation for free space and prove that electromagnetic waves are transverse.
14. Deduce the equation for the propagation of the plane electromagnetic waves in free space. Show that electric and magnetic vectors are normal to each other and to the direction of propagation and are in phase.